AMERICAN AIR POWER AT THE DAWN OF A NEW CENTURY

Benjamin S. Lambeth

Prepared for the United States Navy

Approved for public release; distribution unlimited

The research described in this report was prepared for the United States Navy. The research was conducted in the RAND National Defense Research Institute, a federally funded research and development center sponsored by the Office of the Secretary of Defense, the Joint Staff, the Unified Combatant Commands, the Department of the Navy, the Marine Corps, the defense agencies, and the defense Intelligence Community under Contract DASW01-01-C-0004.

Library of Congress Cataloging-in-Publication Data

Lambeth, Benjamin S.
American carrier air power at the dawn of a new century / Benjamin S. Lambeth.
p. cm.
"MG-404."
Includes bibliographical references.
ISBN 0-8330-3842-7 (pbk. : alk. paper)
1. Aircraft carriers—United States. 2. United States. Navy—Aviation. 3. United States. Marine Corps—Aviation. 4. Afghan War, 2001—Aerial operations, American. 5. Afghan War, 2001—Naval operations, American. 6. War on Terrorism, 2001– 7. Iraq War, 2003-—Aerial operations, American. 8. Iraq War, 2003-—Naval operations, American. I. Title.

V874.3.L43 2005
359.9'4835'0973—dc22

2005023031

The RAND Corporation is a nonprofit research organization providing objective analysis and effective solutions that address the challenges facing the public and private sectors around the world. RAND's publications do not necessarily reflect the opinions of its research clients and sponsors.

Cover design by Peter Soriano

Published 2005 by the RAND Corporation
1776 Main Street, P.O. Box 2138, Santa Monica, CA 90407-2138
1200 South Hayes Street, Arlington, VA 22202-5050
201 North Craig Street, Suite 202, Pittsburgh, PA 15213-1516
RAND URL: http://www.rand.org/
To order RAND documents or to obtain additional information, contact
Distribution Services: Telephone: (310) 451-7002;
Fax: (310) 451-6915; Email: order@rand.org

Preface

This report presents the highlights of the U.S. Navy's carrier air performance during the first two major wars of the 21st century—Operation Enduring Freedom against the Taliban and al Qaeda in Afghanistan in 2001 and 2002 and the subsequent three-week period of major combat in Operation Iraqi Freedom in early 2003 that finally ended the rule of Saddam Hussein. The report also addresses ongoing modernization trends in U.S. carrier air capability. In the first war noted above, U.S. carrier air power substituted almost entirely for land-based theater air forces because of an absence of suitable shore-based forward operating locations for the latter. In the second, six of 12 carriers and their embarked air wings were surged to contribute to the campaign, with a seventh carrier battle group held in reserve in the Western Pacific and an eighth also deployed and available for tasking. The air wings that were embarked in the six committed carriers in the latter campaign flew approximately half the total number of fighter sorties generated altogether by U.S. Central Command. As attested by the performance of naval aviation in both operations, the warfighting potential of today's U.S. carrier strike groups has grown substantially over that of the carrier battle groups that represented the cutting edge of U.S. naval power at the end of the cold war.

The research findings reported herein are the interim results of a larger ongoing study by the author on U.S. carrier air operations and capability improvements since the end of the cold war. They should interest U.S. naval officers and other members of the defense and national security community concerned with the evolving role of U.S. carrier air power in joint and combined operations. The study was sponsored by the Director of Air Warfare (OPNAV N78) in the Office of the Chief of Naval Operations and was conducted in the International Security and Defense Policy Center of the RAND National Defense Research Institute (NDRI). NDRI is a federally funded research and development center sponsored by the Office of the Secretary of Defense, the Joint Staff, the unified Combatant Commands, the Department of the Navy, the Marine Corps, the defense agencies, and the defense Intelligence Community.

For more information on RAND's International Security and Defense Policy Center, contact the Director, James Dobbins. He can be reached by e-mail at James_Dobbins@rand.org; by phone at 703-413-1100, extension 5134; or by mail at the RAND Corporation, 1200 South Hayes Street, Arlington, Virginia 22202-5050. More information about RAND is available at www.rand.org.

Contents

Figures

Summary

The terrorist attacks of September 11, 2001, confronted the Navy, like all of the other U.S. services, with a no-notice call to arms. The sudden demand that they presented for a credible deep-attack capability in the remotest part of Southwest Asia where the United States maintained virtually no access to forward land bases posed an unprecedentedly demanding challenge for naval aviation. Within less than a month after the attacks, the Bush administration and U.S. Central Command (CENTCOM) planned and initiated a campaign to bring down the Taliban theocracy that controlled Afghanistan and that provided safe haven to the terrorist movement that perpetrated the attacks. Code-named Operation Enduring Freedom, that campaign was dominated by air attacks against enemy military assets and personnel, supported by teams of special operations forces (SOF) on the ground working with indigenous Afghan opposition groups to provide U.S. combat aircraft with timely target location, identification, and validation.

Carrier-based Navy and Marine Corps strike fighters operating from stations in the North Arabian Sea substituted almost entirely for Air Force land-based fighter and attack aircraft because of an absence of suitable operating locations close enough to the war zone to make the large-scale use of the latter practicable. Strike missions from the carriers entailed distances to target of 600 nautical miles or more, with an average sortie length of more than four and a half hours. The farthest distance of 750 nautical miles from carrier to targets in northern Afghanistan made for sorties lasting up to ten hours, often with multiple mission tasking.

In all, from the start of hostilities on October 7, 2001, until the end of major offensive operations on March 16, 2002, six carrier battle groups participated in Enduring Freedom. Together, they conducted around-the-clock operations against enemy forces in a landlocked country more than an hour and a half's flight north of the carrier operating areas in the Arabian Sea. Around 80 percent of the carrier-based strike missions dropped ordnance on targets unknown to the aircrews before launch. Of all Navy munitions dropped, 93 percent were either satellite-aided or laser-guided. Each carrier conducted flight operations for roughly 14–16 hours a day, with overlaps as needed to keep an average of three two-plane sections of fighters constantly over Afghanistan for on-call strikes against emerging targets.

This sustained contribution of naval aviation to the campaign (some 72 percent of all combat sorties flown in Enduring Freedom) showed the ability of as many as four carrier battle groups at a time to maintain a sufficient sortie rate to enable a constant armed airborne presence over a landlocked theater more than 400 nautical miles away from the carriers' operating stations in the North Arabian Sea. In so doing, it roundly disconfirmed suggestions voiced by some critics only a few years before that the Navy's carrier force lacked the capability to turn in such a performance. In conducting combat operations throughout the five-month course of major fighting in Enduring Freedom, the participating air wings showed the substantially improved capability that naval strike aviation had acquired since the 1991 Persian Gulf War. The predominant use of precision munitions made the Afghan air war the most precise naval bombing effort in history up to that time.

If Operation Enduring Freedom had been tailor-made for deep-strike carrier air operations, the U.S.-led war against Saddam Hussein in Iraq that followed a year later was no less so, at least with respect to missions launched from the eastern Mediterranean. By the end of the first week of March 2002, as Operation Iraqi Freedom neared, the Navy had two carriers, USS *Theodore Roosevelt* and *Harry S. Truman*, on station in the eastern Mediterranean and three more, USS *Kitty Hawk*, *Constellation*, and *Abraham Lincoln*, deployed in the Persian Gulf. In addition, USS *Nimitz* was en route to the Persian Gulf to relieve *Abraham Lincoln*, which had already been on deployment for an unprecedented nine months.

In all, more than 700 U.S. Navy and Marine Corps aircraft participated in Iraqi Freedom. The average flight operations day aboard each carrier was 16 hours for the first 23 days, after which it ramped down to around 13-14 hours. Each air wing averaged 120-130 sorties a day. Flight deck activity often continued without interruption around the clock for long stretches, since strike aircraft and tankers frequently recovered later than planned as a result of repeated requests for close air support (CAS). As in Operation Enduring Freedom, alert strike packages were launched every day as previously undiscovered targets of interest were identified.

The Iraq war set a new record for close Navy involvement in high-level planning and command of joint air operations. At the operational and tactical levels, the six participating carrier air wings were better integrated into the air-tasking process than ever before, with each wing having full-time representatives in CENTCOM's Combined Air Operations Center to ensure that the wings were assigned appropriate missions. The wings also had ready access to a software package aboard ship that automatically searched the complex daily air operations plan for Navy-pertinent sections, eliminating a need for mission planners and aircrews to study the entire document. Closer cooperation in recent years between the Air Force's and Navy's weapons schools yielded major dividends in improved joint-force interoperability, with the two services working together unprecedentedly well in integrating their respective air operations.

Operations Enduring Freedom and Iraqi Freedom both saw a sustained use of carrier-based air power well beyond littoral reaches. As such, they represented something fundamentally new in the use of naval air power. Unlike previous carrier air applications up to and including Desert Storm a decade before, both wars saw an almost exclusive use of precision-guided munitions by Navy fighters, signaling the advent of a new era in which the principal measure of effectiveness is no longer how many aircraft it takes to neutralize a single target but rather how many aim points can be successfully attacked by a single aircraft. The two wars also saw a pronounced shift from analog to digital network-centric operations, with the Navy's carrier forces increasingly integrated into the digital data stream. In both wars, the performance of the Navy's carrier air wings offered a strong validation of the final maturation of U.S. carrier air power after more than a decade of programmatic setbacks and drift in the wake of the cold war's end.

Before the terrorist attacks of September 11, the Navy's global presence posture had been enabled by a highly routinized sequence of maintenance, training, and unit and ship certification aimed at meeting scheduled deployment dates that were all but carved in stone. The sudden demands levied on the Navy by the terrorist attacks, however, changed that pattern of operations irretrievably. Recognizing that the new demands of an open-ended global war on terror meant a need for more responsive naval forces able to sustain a higher level of readiness, the Chief of Naval Operations (CNO) in March 2003 announced a need for the Navy to develop a new Fleet Response Concept (FRC) aimed at providing an enhanced carrier surge capability.

That initiative was put into effect on the eve of Operation Iraqi Freedom, which offered a timely opportunity to test the concept under fire. As the war neared, the Navy had eight carrier battle groups deployed, including USS *Carl Vinson* in the Western Pacific monitoring North Korea and China during the final countdown. Five of those eight battle groups and air wings had participated in Operation Enduring Freedom just a year before. With five battle groups on station and committed to the impending war, a sixth en route to the war zone as a timely replacement for one of those five, a seventh also forward-deployed and holding in ready reserve, and yet an eighth carrier at sea and available for tasking, a full 80 percent of the Navy's carrier-based striking power was deployed and combat-ready. With that demonstrated performance having validated the FRC proposal, the CNO in the early aftermath of Iraqi Freedom approved it and directed its implementation as the Fleet Response Plan (FRP).

The FRP seeks to increase the efficiency of maintenance and training processes and procedures so as to heighten overall carrier availability and readiness and to increase the carrier force's speed of employment. It envisages the augmentation of deployed carrier battle groups with surgeable battle groups ready for deployment and combatant-commander tasking, thereby yielding increased overall force employability and earlier commitment of carrier striking power. More specifically, it aims to provide combatant commanders with what has come to be characterized as "six-plus-two" ready carrier strike groups (CSGs). The "six" refers to deployable CSGs that can respond almost immediately to tasking, wherever they may be in their respective training schedules, in varying amounts of time up to 30 days. The remaining two represent near-combat-ready CSGs that can deploy as needed on a more accelerated schedule than before, normally within around 90 days. That will constitute a larger overall naval air force complement able to respond to tasking, as opposed to a smaller forward-deployed force fielded primarily to meet "presence" requirements.

With respect to planned force modernization, the Navy's nuclear-powered *Nimitz*-class aircraft carriers have provided the nation with effective service for more than 30 years. The design for those carriers was completed during the 1960s. Since then, the carrier force has not undergone an aggressive effort to fold cutting-edge technology into the design of follow-on platforms. In light of these considerations, and prompted by growing concern that the continued absence of any significant progress in U.S. carrier design was inhibiting operational capability improvements and the incorporation of new technologies, the Navy in 1993 commissioned a future sea-based air platforms working group to explore operational requirements, available systems and technologies, and needed R&D initiatives for defining and developing the next generation of carriers. That initiative eventually resulted in the establishment of the Future Aircraft Carriers (CVX) program.

Largely on the strength of subsequent analytical assessments and findings, the Defense Acquisition Board (DAB) in June 2000 approved the Navy's proposed plan to pursue a follow-on to the

Nimitz-class carrier that will be a large-deck, nuclear-powered ship that was then designated CVNX. The DAB's consensus was that large-deck carriers were the preferable alternative for a broad range of reasons having to do with operational effectiveness and flexibility. Once commissioned in 2015, as its current development schedule projects, the first of the Navy's next-generation carriers, now called CVN-21, will feature such radical innovations as an advanced reactor and electrification of all auxiliary systems, which will increase the ship's electrical power-generation capability to three times that of the *Nimitz* design and will also replace steam and hydraulic piping throughout the ship. In addition, four electromagnetic aircraft launch catapults will replace the earlier-generation steam catapults. CVN-21 will have a more efficient flight deck and advanced arresting gear for aircraft recoveries. An abiding hallmark of its many design goals is the provision of an adaptable infrastructure that will allow the incorporation of new capabilities as they develop. These measures will greatly reduce life-cycle costs over the new carrier's planned service life.

Among the many gains that have been registered in the leverage of U.S. carrier air power over the past decade have been a proven ability to surge a large number of CSGs (as many as eight out of 12 carriers and ten air wings) and to keep them on station for the duration of a major campaign; to attack multiple aim points with consistently high accuracy on each combat sortie around the clock irrespective of weather; and—with the help of nonorganic tanking support—conduct deep-strike missions well beyond littoral reaches and remain on station for hours, if need be, in providing on-call interdiction and CAS. These are new capabilities that would have been all but unthinkable during the final years of the cold war, even when the Navy maintained 15 active carrier battle groups.

In the decade ahead, this process of evolutionary improvement in naval strike warfare will continue unfolding in a way that promises revolutionary advances in the potential of U.S. carrier air power. In particular, the immediate years ahead will see a further sharpening of the edge of the F/A-18E/F Super Hornet, a successor generation of naval electronic warfare aircraft in the EA-18G, the introduction of the E-2D Advanced Hawkeye offering significantly increased airborne surveillance and battle-management capabilities, the Navy's long-overdue acquisition of an all-aspect stealth platform with the pending introduction of the F-35C Joint Strike Fighter, and a leaner yet more efficient and capable carrier air-wing force structure.

The Navy also is evolving from being a platform-centric to a network-centric force. A recent CNO initiative called FORCENet aims to tie together naval, joint, national, and ultimately coalition information grids to achieve an unprecedented level of battlespace awareness and knowledge management at all levels. This initiative will allow improved situation awareness, quicker battle-damage assessment, and real-time target reattack decisionmaking. It also will provide a common operating picture up and down the chain of command, from the most senior leadership all the way into the cockpits of individual shooters at the tactical level.

In sum, Operations Enduring Freedom and Iraqi Freedom showed that the Navy's carriers no longer operate as individual and autonomous air-wing platforms but rather as a surged and massed force capable of generating and sustaining however many consistently effective sorties over time that an air component commander may need to meet his assigned campaign goals. Unlike the relatively short-range sorties flown during the largely demonstrative and punitive strikes launched into Lebanon in 1983 and against Libya in 1986 and in such subsequent contingency responses during the 1990s as Operations Deliberate Force, Desert Fox, and Allied Force, these were multicycle missions that lasted for as long as ten hours and that ranged deep beyond littoral reaches into the heart of Afghanistan and Iraq, the first of which was landlocked in the most remote part of Central Asia.

Today, carrier aviation is not only a natural concomitant of the nation's status as the world's sole surviving superpower, it also is the one outstanding feature that distinguishes the U.S. Navy unequivocally from all other naval forces around the world. The *Nimitz*-class carrier has often been described as four and a half acres of sovereign U.S. territory that can go anywhere the nation's leaders may desire to send it without needing a permission slip. For years, that characterization was dismissed by critics of carrier air power as a mere slogan that overlooked the fact that a carrier can be in only one place at a time, irrespective of where a need for it might suddenly arise. That criticism may have had merit throughout most of the cold war, when the Navy typically kept only two or three carrier battle groups deployed at any time, while the others and their attached air wings remained at home in various states of maintenance and requalification training that rendered them undeployable on short notice. That is no longer the case, however, in today's world of constant carrier surge capability under the FRP. When U.S. naval aviation celebrates its 100th anniversary in 2011, carrier air power's classic roles and missions will not have changed greatly from those of the 20th century. Yet the nation's carrier strike groups will have taken on a substantial qualitative improvement in their overall combat leverage with the completion of the Super Hornet acquisition, the advent of the EA-18 and F-35C, and the prospective introduction of unmanned surveillance and strike aircraft into the Navy's carrier air-wing complements.

Acknowledgments

This report has benefited from the helpful discussions I had during the course of preparing it with then–Vice Admiral Timothy Keating, Director, Joint Staff; then–Vice Admiral John Nathman, Commander, Naval Air Force, U.S. Pacific Fleet; then–Rear Admiral Lewis Crenshaw, Director, Assessments Division, OPNAV N81; then–Rear Admiral Evan Chanik, Director, Programming Division, OPNAV N80; Rear Admiral John Cryer, III, Commander, Naval Space Command; then–Rear Admiral Mark Fitzgerald and Rear Admiral Thomas Kilcline, successive Directors of Air Warfare, OPNAV N78; Rear Admiral Richard Gallagher, Commander, Strike Force Training, U.S. Atlantic Fleet; Rear Admiral William Gortney, U.S. Fleet Forces Command; Rear Admiral Thomas Zelibor, Director, Space, Information Warfare and Command and Control, OPNAV N61; then–Rear Admiral James Zortman, Commander, Naval Air Force, U.S. Atlantic Fleet; then–Rear Admiral (select) James Winnefeld, Jr., Executive Assistant to the Vice Chief of Naval Operations; Rear Admiral Richard Wren, OPNAV N78; Captain Martin Erdossy, OPNAV N78; Captain Robert Nelson, OPNAV N3/5 (Deep Blue); Captain Janice Hamby, OPNAV N70; Captain Charles Wright, Director for Naval Aviation Systems, Office of the Secretary of Defense (Operational Test and Evaluation); then–Commander Calvin Craig, OPNAV N81; Captain Kenneth Neubauer and Lieutenant Commander Nicholas Dienna, both former Navy Executive Fellows at RAND; Commander Andrew Lewis, Executive Assistant to the Commander, Naval Air Force, U.S. Atlantic Fleet; Malcolm Taylor,

Principal Assistant for Air Warfare Plans, Analysis and Assessments, OPNAV N783B; Lieutenant Commander Scott Moran, OPNAV N780C3B; and my RAND colleagues John Birkler, James Dobbins, Richard Hundley, and Rear Admiral Francis Lacroix, USN (Ret.). I thank all of the above for their good insights and enthusiastic support of this effort. I am especially grateful to Admirals Chanik, Crenshaw, Fitzgerald, Gallagher, and Kilcline; Rear Admiral Matthew Moffit, Commander, Naval Strike and Air Warfare Center; Captain Hamby; Captain Wright; Captain James McDonell, USN (Ret.), former commanding officer of USS *John C. Stennis* (CVN-75); Commander James Bynum, Executive Assistant to the Commander, Naval Air Force, U.S. Pacific Fleet; Commander Craig; and Malcolm Taylor for their critique of all or parts of an earlier draft, as well as to Christine Fox at the Center for Naval Analyses and Captain Nelson in Deep Blue for providing me some useful documentary materials. I also thank my RAND colleague Irv Blickstein and Adam Siegel of the Northrop Grumman Analysis Center for their expert peer reviews of the final manuscript and Chris Jantsch, Naval Air Systems Command, for his help in providing the images for the cover art.

I might note in passing that the report has been additionally enriched by a number of opportunities I was privileged to have in connection with RAND work in earlier years to gain a broad range of first-hand exposure to the world of naval air operations in the training environment. Those opportunities included six adversary training sorties in the TA-4J with VF-126 out of NAS Miramar, California, and NAS Fallon, Nevada, in 1980; three F-5F syllabus sorties with Navy Fighter Weapons School (TOPGUN) at Miramar in connection with my attending the first week of the TOPGUN course in 1980; four F-14A sorties, including two arrested landings in USS *Kitty Hawk* (CV-63), with VF-1 out of Miramar in 1983; 15 basic air-to-air and surface-attack training sorties in the TA-4J with VT-24 and an advanced aircraft-handling sortie in a T-2C with VT-26 out of NAS Beeville, Texas, in 1985; a TA-7C sortie with the Naval Strike Warfare Center at Fallon in 1986; four air-to-air sorties in a Navy F/A-18B from VFA-125 out of NAS Lemoore, California, during the four-day Defensive Anti-Air Warfare Phase of the Weapons and Tactics Instructor course offered quarterly by Marine Aviation Weapons and Tactics Squadron One at MCAS Yuma, Arizona, in 1986; an EA-6B mission orientation flight with VMAQ-2 at MCAS Cherry Point, North Carolina, in 1987; and an F-14A+ sortie with VF-24 out of Miramar in 1990. I remain deeply indebted to all in the Navy's and Marine Corps' flight approval chains for civilians during that ten-year window of time who made it possible for me to acquire this invaluable hands-on experience. It has, at long last, produced a direct return on investment by the sea services nearly half a generation later in a way that no one could have anticipated at the time.

Acronyms

AAA	Antiaircraft Artillery
AARGM	Advanced Anti-Radiation Guided Missile
ABCCC	Airborne Command and Control Center
AESA	Active Electronically Scanned Array
AFB	Air Force Base
AFSB	Afloat Forward Staging Base
AFSOC	Air Force Special Operations Command
AGM	Air-to-Ground Missile
AIM	Air Intercept Missile
AMRAAM	Advanced Medium-Range Air-to-Air Missile
AOR	Area of Responsibility
ASW	Antisubmarine Warfare
ATFLIR	Advanced Tactical Forward-Looking Infrared
ATO	Air Tasking Order
AWACS	Airborne Warning and Control System
BAMS	Broad-Area Maritime Surveillance
BDA	Battle Damage Assessment
BLU	Bomb Live Unit
CAOC	Combined Air Operations Center
CAP	Combat Air Patrol
CAS	Close Air Support

CBU	Cluster Bomb Unit
CENTAF	Central Command Air Forces
CENTCOM	Central Command
CFACC	Combined Force Air Component Commander
CFMCC	Combined Force Maritime Component Commander
CJTF	Combined Joint Task Force
CNO	Chief of Naval Operations
CSG	Carrier Strike Group
CTOL	Conventional Takeoff and Land
CV	Carrier
CVN	Nuclear Carrier
CVNX	Nuclear Carrier Experimental
CVW	Carrier Air Wing
CVX	Carrier Experimental
DAB	Defense Acquisition Board
DMPI	Desired Mean Point of Impact
DPG	Defense Planning Guidance
ELINT	Electronic Intelligence
EMALS	Electromagnetic Aircraft Launch System
ESG	Expeditionary Strike Group
EW	Electronic Warfare
FAC	Forward Air Controller
FAC-A	Airborne Forward Air Controller
FLIR	Forward-Looking Infrared
FMC	Fully Mission-Capable
FRC	Fleet Response Concept
FRP	Fleet Response Plan
FSCL	Fire Support Coordination Line
FTI	Fast Tactical Imagery

FYDP	Future Years Defense Program
GBU	Guided Bomb Unit
GPS	Global Positioning System
HARM	High-Speed Anti-Radiation Missile
HMCS	Helmet-Mounted Cueing System
HUD	Head-Up Display
IADS	Integrated Air Defense System
ICAP	Improved Capability
ISR	Intelligence, Surveillance, and Reconnaissance
JAST	Joint Advanced Strike Technology
JCS	Joint Chiefs of Staff
JDAM	Joint Direct Attack Munition
JFN	Joint Fires Network
JSOW	Joint Standoff Weapon
J-UCAS	Joint Unmanned Combat Aerial System
LANTIRN	Low-Altitude Navigation and Targeting Infrared for Night
LD/HD	Low-Density/High-Demand
LGB	Laser-Guided Bomb
LSO	Landing Signals Officer
MAGTF	Marine Air-Ground Task Force
MIDS	Multifunction Information Distribution System
MMA	Multimission Maritime Aircraft
MRC	Major Regional Contingency
NAS	Naval Air Station
NEO	Noncombatant Evacuation Operation
NRAC	Naval Research Advisory Committee
NSAWC	Naval Strike and Air Warfare Center
OPNAV	Office of the Chief of Naval Operations

OSD	Office of the Secretary of Defense
PGM	Precision-Guided Munition
RAF	Royal Air Force
ROE	Rules of Engagement
RIO	Radar Intercept Officer
SAM	Surface-to-Air Missile
SAR	Synthetic Aperture Radar
SEAD	Suppression of Enemy Air Defenses
SFWT	Strike Fighter Weapons and Tactics
SHARP	Shared Reconnaissance Pod
SIPRNET	Secure Internet Protocol Router Network
SLAM	Standoff Land Attack Missile
SOF	Special Operations Forces
SPIN	Special Instruction
STOVL	Short Takeoff/Vertical Landing
TACP	Tactical Air Control Party
TARPS	Tactical Air Reconnaissance Pod System
TCS	Television Camera System
TLAM	Tomahawk Land-Attack Missile
TST	Time-Sensitive Targeting
TTP	Tactics, Techniques, and Procedures
T3	Tomcat Tactical Targeting
UAV	Unmanned Aerial Vehicle
UCAV	Unmanned Combat Aerial Vehicle
UHF	Ultra High Frequency
USS	United States Ship
VF	Navy Fighter Squadron
VFA	Navy Fighter/Attack Squadron

CHAPTER ONE

Introduction

Throughout most of the cold-war years after American combat involvement in Vietnam ended in 1973, the U.S. Navy's aircraft carriers figured most prominently in an offensive sea-control strategy that was directed mainly against Soviet naval forces, including long-range and highly capable shore-based naval air forces, for potential open-ocean (or "blue-water") engagements around the world in case of major war. For lesser contingencies, the principal intended use of the Navy's carrier battle groups was in providing forward "presence" to symbolize American military power and global commitment. When it came to actual force employment, however, U.S. carrier-based aviation was typically used only in occasional one-shot demonstrative applications against targets located in fairly close-in littoral areas, such as the carrier-launched air strikes against Syrian forces in Lebanon in 1983 and Operation El Dorado Canyon against Libya's Moammar Ghaddafi in 1986.

Iraq's sudden and unexpected invasion of Kuwait in August 1990, however, presented American carrier air power not only with its first crisis of the post-cold-war era, but also with a novel set of challenges that amounted to a wake-up call for the Navy as it confronted the unfamiliar demands of an emerging new era. Over the course of the six-week Persian Gulf War that began five and a half months later, the Navy's carrier force found itself obliged to make a multitude of adjustments during that war. Few of the challenges that were levied on naval aviation by that U.S.-led offensive, code-named

Operation Desert Storm, bore much resemblance to the planning assumptions that underlay the Navy's Maritime Strategy that had been created to accommodate a very different set of operational concerns during the early 1980s.

Simply put, Desert Storm in no way resembled the open-ocean showdowns between opposing high-technology forces that the Navy had planned and prepared for throughout the preceding two decades. Instead, it was replete with the sort of challenges that were unique to littoral operations. To begin with, there were no significant enemy surface naval forces or air threat to challenge the Navy's six carrier battle groups that participated in that war. Moreover, throughout the course of the brief campaign and the five-month buildup of forces in the region that preceded it, the Navy did not operate independently, as was its habit throughout most of the cold war, but rather in shared operating areas with the U.S. Air Force and Army. Because of the Navy's lack of a compatible command and control system, the daily Air Tasking Order (ATO) generated by U.S. Central Command's (CENTCOM's) Air Force-dominated Combined Air Operations Center (CAOC) in Saudi Arabia had to be placed aboard two Navy S-3 aircraft in hard copy each day and flown to the participating carriers so that the next day's air-wing flight schedules could be written.

Furthermore, the naval air capabilities that had been fielded and fine-tuned for open-ocean engagements, such as the long-range AIM-54 Phoenix air-to-air missile carried by the F-14 fleet defense fighter, were of little relevance to the allied coalition's combat needs.[1] Navy F-14s were not assigned to the choicest combat air patrol (CAP) stations in Desert Storm because, having been equipped for the less crowded outer air battle in defense of the carrier battle group, they lacked the redundant onboard target recognition systems that CENTCOM's rules of engagement required for the denser and more confused air operations environment over Iraq. As for the Navy's

[1] Edward J. Marolda and Robert J. Schneider, Jr., *Sword and Shield: The United States Navy and the Persian Gulf War*, Annapolis, Md.: Naval Institute Press, 1998, pp. 180–181. See also James A. Winnefeld and Dana J. Johnson, *Joint Air Operations: Pursuit of Unity in Command and Control, 1942–1991*, Annapolis, Md.: Naval Institute Press, 1993, p. 115.

other habit patterns and items of equipment developed for open-ocean engagements, such as fire-and-forget Harpoon antiship missiles, level-of-effort ordnance planning, and decentralized command and control, all were, in the words of the former Vice Chairman of the Joint Chiefs of Staff (JCS), Admiral William Owens, "either ruled out by the context of the battle or were ineffective in the confined littoral arena and the environmental complexities of the sea-land interface."[2] U.S. naval aviation performed admirably in Desert Storm only because of its inherent professionalism and adaptability, not because its doctrine and weapons complement were appropriate to the situation.

The Navy, however, soon moved out smartly to make the needed readjustments to the emerging post-cold-war era beginning in the early aftermath of the Persian Gulf War. For example, in response to identified shortcomings that were spotlighted by its Desert Storm experience, the Navy substantially upgraded its precision-strike capability by fielding new systems and adding improvements to existing platforms that gave carrier aviation a degree of flexibility that it had lacked throughout Desert Storm. First, it took determined steps to convert its F-14 fleet defense fighter from a single-mission air-to-air platform into a true multimission aircraft through the incorporation of the Air Force–developed LANTIRN infrared targeting system that allowed the aircraft to deliver laser-guided bombs with consistently high accuracy both day and night.[3] Starting in 1997, the Navy ultimately modified 222 F-14s to carry the LANTIRN system, giving the aircraft a precision deep-attack capability that put it in the same league as the Air Force's F-15E Strike Eagle. In the process, the F-14 relinquished much of its former strike escort role and left that to the F/A-18 with the AIM-120 advanced medium-range air-to-air missile (AMRAAM) as the Tomcat was transformed, in effect, into the deep

---

[2] Then–Vice Admiral William Owens, USN, "The Quest for Consensus," *Proceedings*, May 1994, p. 68.

[3] LANTIRN is an acronym for low-altitude navigation and targeting infrared for night.

precision-attack A-6 of old with its much-improved LANTIRN targeting capability.

To correct yet another deficiency highlighted by the Desert Storm experience, naval aviation also undertook measures to improve its command, control, and communications arrangements so that it could operate more freely with other joint air assets within the framework of an ATO. Those measures most notably included the gaining of a long-needed ability to receive the daily ATO aboard ship electronically. In addition, the Navy made provisions for a more flexible mix of aircraft in a carrier air wing, which could now be tailored to meet the specific needs of a joint force commander. The new look of naval aviation also featured a closer integration of Navy and Marine Corps air assets that went well beyond the mere "coordination" that had long been the rule hitherto. That initiative resulted in a greater synergy of forces occasioned by physically blending Marine F/A-18 strike-fighter squadrons into Navy carrier air wings as a matter of standard practice.

Finally, there was an emergent Navy acceptance of the value of strategic air campaigns and the idea that naval air forces must be more influential players in them. As Admiral Owens noted as early as 1995, "the issue facing the nation's naval forces is not whether strategic bombardment theory is absolutely correct; it is how best to contribute to successful strategic bombardment campaigns."[4] The Navy leadership freely acknowledged that its shortfall in precision-guided munitions (PGMs) had limited the effectiveness of naval air power in Desert Storm, a gap that it subsequently narrowed through the improvements to the F-14 noted above and by equipping more Navy and Marine Corps F/A-18s with the ability to fire the AGM-84 standoff land attack missile (SLAM) and to drop the satellite-aided GBU-31 2,000 lb joint direct attack munition (JDAM).

Despite these and related readjustments, however, naval aviation was by no means out of the woods just yet. On the contrary, the ending of the cold war, which occurred more or less concurrently

[4] Admiral William A. Owens, USN (Ret.), *High Seas: The Naval Passage to an Uncharted World*, Annapolis, Md.: Naval Institute Press, 1995, p. 96.

with the successful conclusion of Desert Storm, further accelerated an already ongoing decline in U.S. defense spending, begun late during the Reagan years and continued by the first Bush administration, to a lower level in constant dollars and percentage of gross domestic product than any experienced by the United States since before the outbreak of the Korean War. Emblematic of this emergent trend was the cancellation of the troubled A-12 stealth attack aircraft program in 1991 by then–Secretary of Defense Dick Cheney on grounds of uncontrolled cost escalation and reduced operational need. That aircraft had been intended to replace the venerable A-6 medium bomber and, in the process, to bring the Navy into the stealth era in a major way.

For the Navy, the post-cold-war U.S. force drawdown that ensued included a loss of three out of 15 deployable carrier battle groups and a concomitant decline in the number of authorized strike-capable aircraft by almost half. As the Chief of Naval Operations (CNO) during the early aftermath of that drawdown, Admiral Jay Johnson, described its impact, "if we have a two-carrier presence in the Gulf, it means we have a zero presence somewhere else."[5] Granted, part of this force reduction simply reflected the growing obsolescence of certain older aircraft that had been in the Navy's inventory for more than three decades and were long overdue to be retired. For instance, the workhorse A-6 medium-attack aircraft, the last of which was retired from the fleet in 1997, had been in service with the nation's carrier force since the early 1960s. Nevertheless, the Navy, like all of the other U.S. armed services, entered the last decade of the 20th century being asked to do ever more with ever less.

As it suffered one major aircraft program cancellation after another during the early and mid-1990s (with the stealthy AX and A/FX going by the boards in close succession after the A-12's demise), naval aviation also took multiple broadside hits in the increasingly competitive and combative interservice roles and resources arena. One common criticism of carrier air power levied by Air Force proponents during the mid-1990s charged that "for anything other

[5] Bradley Graham, "U.S. Military Feels Strain of Buildup," *Washington Post*, February 5, 1998.

than a one-time show-of-force strike . . . a carrier battle group would be badly handicapped in comparison with a wing of B-2s, even if the battle group was on hand and the bomber wing staged initially from the U.S."[6] Another pro-Air Force detractor of sea-based air power wrote as recently as 1999 that carrier air effectiveness had been falsely inflated to "mythic proportions" by its most outspoken proponents, particularly with respect to alleged claims that carriers can operate without access to land bases and can "carry out sustained strikes against targets several hundred miles inland." This critic cited the Navy's much-heralded Surge 97 exercise's short-sortie evolution as alleged proof that "targets more than 500 miles from the carrier would prove to be out of reach," concluding from this that the scenario had "reflected a blue-water, ocean-control legacy" rather than "a realistic littoral scenario."[7] As if to bear this charge out, throughout the later post-cold-war years that followed the 1991 Persian Gulf War, the involvement of the Navy's carrier air wings in such regional contingency responses as Operations Deliberate Force and Allied Force in the Balkans and Operations Southern Watch and Desert Fox over Iraq mainly entailed relatively low-intensity operations conducted within fairly easy reach of their assigned targets.

The dawn of the 21st century, however, heralded the start of a fundamentally new era for U.S. carrier-based aviation. The terrorist attacks against the United States on September 11, 2001, portended a change of major proportions in the long-familiar pattern of U.S. carrier air operations. Those attacks imposed a demand for a credible deep-strike capability in the remotest part of Southwest Asia where the United States maintained virtually no access for forward land-based air operations. That demand presented a new and unique challenge for the nation's carrier force. Less than a month after the attacks

---

[6] Colonel Brian E. Wages, USAF (Ret.), "Circle the Carriers: Why Does 'Virtual Presence' Scare the Navy," *Armed Forces Journal International*, July 1995, p. 28.

[7] Rebecca Grant, "The Carrier Myth," *Air Force Magazine*, March 1999, p. 26. The most complete account of this exercise, which freely admits some of the exercise's necessary artificialities, remains Angelyn Jewell, Maureen A. Wigge, and others, *USS Nimitz and Carrier Air Wing Nine Surge Demonstration*, Alexandria, Va.: Center for Naval Analyses, CRM 97-111.10, April 1998.

perpetrated by Osama bin Laden and his al Qaeda terrorist organization, the nation found itself at war against al Qaeda's main base structure in Afghanistan and against the ruling Taliban theocracy that had provided it safe haven. In that response, code-named Operation Enduring Freedom, carrier-based Navy and Marine Corps strike fighters operating from stations in the North Arabian Sea substituted almost entirely for Air Force land-based fighter and attack aircraft because of an absence of suitable operating locations close enough to the war zone to make the large-scale use of the latter practicable. In the process, the carrier air wings that deployed to the region generated the vast majority of the strike-fighter sorties that were flown throughout the war.

Barely more than a year later, the Navy's carrier force again played a pivotal role when five battle groups and their embarked air wings took up stations (three in the Arabian Gulf and two in the eastern Mediterranean Sea) in preparation for Operation Iraqi Freedom, which commenced on March 19, 2003. Over the course of that three-week period of major combat, the five carriers—with a sixth en route to the region to replace one, a seventh held in reserve in the Western Pacific, and an eighth also deployed and available for tasking—conducted around-the-clock operations against Saddam Hussein's forces in Iraq. With the support of nonorganic U.S. Air Force and British Royal Air Force (RAF) long-range tankers to provide multiple inflight refuelings, combat aircraft from the two carriers operating in the eastern Mediterranean flew repeated deep-strike missions that entailed durations of as long as ten hours, in some cases.

Both of these major carrier air operations in close succession saw a sustained use of U.S. naval air assets well beyond littoral reaches. As such, they represented something never before experienced in the evolution of American carrier-based air power. In addition, the two wars saw naval aviation more fully represented than ever before throughout CENTCOM's CAOC at Prince Sultan Air Base in Saudi Arabia, which was the nerve center for all air operations in both cases. They also saw naval aviation fully integrated into the joint and combined air operations that largely enabled the successful outcomes in each case.

Unlike past naval air applications up to and including the 1991 Persian Gulf War a decade before, both wars saw an almost exclusive use of precision-guided munitions by Navy strike fighters, signaling the advent of a new era in which the principal measure of effectiveness is now no longer how many aircraft it might take to destroy a single target but rather how many target aim points can be successfully attacked by a single aircraft. The two wars also saw a pronounced shift from analog to digital network-centric operations, with the Navy's carrier forces increasingly integrated into the digital data stream. None of these achievements would have been possible at the height of the cold war, when U.S. naval aviation was configured differently and oriented toward meeting a very different spectrum of challenges. In both wars, the performance of the Navy's carrier battle groups and air wings offered a resounding validation of the final maturation of U.S. carrier-based air power after more than a decade of setbacks and programmatic drift in the wake of the cold war's end.

CHAPTER TWO

Carrier Air over Afghanistan

The attacks planned and executed against the United States by Osama bin Laden and his al Qaeda terrorist organization on September 11, 2001, confronted the Navy, like all the other armed services, with a no-notice call to arms. Earlier throughout the years that followed the 1991 Persian Gulf War, the Navy's carrier battle groups had taken part in numerous contingency-response operations that served to further hone the edge of the nation's carrier air forces. For the most part, however, those operations involved fairly short distances to target and few significant stresses on carrier aviation. In sharp contrast, the looming demand for a credible deep-attack capability into the remotest part of Southwest Asia where the United States maintained virtually no access to forward land bases confronted the Navy's carrier force with a uniquely demanding challenge.

At the time the attacks occurred, the aircraft carriers USS *George Washington* (CVN-73) and *John F. Kennedy* (CV-67) were engaged in predeployment workups off the East Coast of the United States. *John C. Stennis* (CVN-74) and *Constellation* (CV-64) were similarly preparing for deployment off the California coast. *Kitty Hawk* (CV-63) was at dockside in her home port of Yokosuka, Japan. *Enterprise* (CVN-65) was outbound from the Southwest Asian area of operations off the coast of Yemen heading for home as she neared the end of a six-month deployment to the Persian Gulf. *Carl Vinson* (CVN-70) was inbound to CENTCOM's area of responsibility (AOR) off the southern tip of India to relieve *Enterprise*.

These ships and numerous others were ordered to their highest state of readiness in the immediate aftermath of the attacks. The Department of Defense and the carrier battle group commanders also initiated moves to update contingency plans for naval strike operations in the most likely areas of possible U.S. combat involvement worldwide. Rightly deducing that his ship's presence would soon be needed in the Afghanistan area of operations, the commanding officer of *Enterprise* immediately turned his ship around upon learning of the terrorist attacks and was subsequently ordered to remain in the region for an indefinite length of time.[1]

At the same time, *Carl Vinson* was rerouted from her previously assigned operating area to join *Enterprise* in the North Arabian Sea. That doubled the normal number of carrier air wings ready for tasking in that portion of CENTCOM's AOR. *Theodore Roosevelt*, with her battle group of around a dozen ships and a three-ship Marine Corps amphibious ready group, was slated to sail from Norfolk the week of September 19. Once she was under way, the Navy would have five of its 12 carriers headed toward CENTCOM's AOR simultaneously.[2]

Concurrently, *Kitty Hawk* departed Yokosuka without her full air wing aboard to provide what later came to be referred to as a sea-based "lily pad" from which U.S. special operations forces (SOF) teams would be staged into Afghanistan. To free up her flight and hangar decks to make room for a variety of SOF helicopters, *Kitty Hawk* carried only a small presence of eight F/A-18 strike fighters from her normal air-wing complement of more than 50 combat aircraft, primarily to provide an air defense shield for the battle group. She would not arrive on station in the AOR until October 13. By October 1, however, *Carl Vinson* and *Enterprise* were in position to commence strike operations, with *Theodore Roosevelt* expected to be ready to join them in the North Arabian Sea within a week. By this

[1] Greg Jaffe, "U.S. Armed Forces Are Put on the Highest State of Alert," *Wall Street Journal*, September 12, 2001.

[2] Christian Bohmfalk and Jonathan Block, "Roosevelt Carrier Battle Group Scheduled to Deploy Wednesday," *Inside the Navy*, September 17, 2001.

time, the overall number of U.S. aircraft in the region had grown to between 400 and 500, including 75 on each of the Navy's three carriers on station.

Within less than a month after the al Qaeda terrorists flew the airliners that they had hijacked into the twin towers of the World Trade Center in New York and into the Pentagon just south of Washington, D.C., the administration of President George W. Bush and the commander and staff of CENTCOM organized, planned, and initiated a joint and combined campaign to bring down the Taliban theocracy that controlled Afghanistan and that had provided bin Laden and his terrorist operation safe haven there since 1998. Code-named Operation Enduring Freedom, that campaign would be dominated by air attacks against Taliban and al Qaeda military assets and personnel, supported by SOF teams on the ground working with indigenous Afghan opposition groups to provide allied strike aircraft with timely target location, identification, and validation.

To be sure, Air Force heavy bombers also played a prominent part in the air campaign, flying from the British island base of Diego Garcia in the Indian Ocean and, in the case of the B-2 stealth bomber (which flew six missions against Taliban air defenses during the campaign's first two nights), all the way from Whiteman AFB, Missouri, and back. Indeed, Air Force bombers dropped nearly three-quarters of all the satellite-aided JDAMs delivered throughout the war. Air Force F-15E and F-16 fighters also contributed materially to strike operations, albeit in far smaller numbers and only after the tenth day once the needed forward basing arrangements had been secured, by flying long-duration combat sorties into Afghanistan from several friendly countries in the Persian Gulf. Nevertheless, as a part of the joint force, carrier-based Navy and Marine Corps strike fighters operating from stations in the North Arabian Sea substituted almost entirely for Air Force land-based fighter and attack aircraft because of an absence of suitable operating locations close enough to the war zone to make the large-scale use of the latter practicable. In so doing, the Navy's carrier air wings that were committed to the campaign provided CENTCOM with a crucial contribution to combat operations throughout the war.

Naval Aviation Goes to War

Operation Enduring Freedom began under clear skies during the evening of October 7, 2001, with air attacks against targets in Kabul and in the southern Taliban stronghold area of Kandahar. Beginning three days before the onset of actual combat, F-14s configured with the Tactical Air Reconnaissance Pod System (TARPS) flew armed reconnaissance missions over major areas of interest in Afghanistan. That application was a significant contribution by the F-14 as both a legacy platform and the sole organic tactical reconnaissance capability left available to the battle group commander. Once the bombing was under way, the Navy's initial targets consisted of Taliban airfields, air defense positions, command and control nodes, and al Qaeda terrorist training camps.

The opening-night attacks were carried out by 25 F-14 and F/A-18 strike fighters launched from *Enterprise* and *Carl Vinson* operating in the North Arabian Sea, along with five U.S. Air Force B-1B, ten B-52, and two B-2 bombers. These attack aircraft were supported by accompanying F-14 and F/A-18 fighter sweeps, as well as by surveillance and aircraft flow control provided by E-2Cs and by radar and communications jamming provided by EA-6B Prowlers.[3] During these operations, the Navy's EA-6Bs played a new role. In the past, they had focused mainly on jamming enemy surface-to-air missile (SAM) radars. In Enduring Freedom, they jammed Taliban radars during the first days until enemy air defenses had been largely neutralized but then refocused for the first time on jamming enemy ground communications.[4]

Those operations were supported by an elaborate inflight refueling scheme, with carrier-based S-3 tankers orbiting off the coast of Pakistan to top off inbound Navy strike aircraft just before the latter proceeded to their holding stations over Afghanistan. Air Force KC-

---

[3] Thomas E. Ricks and Vernon Loeb, "Initial Aim Is Hitting Taliban Defenses," *Washington Post*, October 8, 2001.

[4] Randy Woods, "Prowler, Hawkeye Pilots See Roles Expanding in Enduring Freedom," *Inside the Navy*, May 6, 2002.

135 and KC-10 tankers, supplemented by RAF Tristars and VC-10s, orbited farther north to refuel the strike aircraft again as mission needs required before the latter returned to their ships.[5] Strike missions from the carriers entailed distances to target of 600 nautical miles or more, with an average sortie length of more than four and a half hours and a minimum of two inflight refuelings each way to complete the mission.[6]

Throughout the first five days, Navy fighters dropped 240 JDAMs and laser-guided bombs (LGBs) altogether, as well as one I-2000 BLU-109 hard-structure munition.[7] A week later, in three consecutive days of the war's heaviest bombing to date, allied aircraft attacked a dozen target sets, including Taliban airfields, antiaircraft artillery (AAA) positions, armored vehicles, ammunition dumps, and terrorist training camps. Those attacks involved some 90 Navy and Marine Corps fighters operating from all three air wings that were by then on station aboard *Enterprise*, *Carl Vinson*, and *Theodore Roosevelt*.[8] The farthest distance of 750 nautical miles from carrier to targets in northern Afghanistan made for sorties lasting, on occasion, as long as ten hours, often with multiple mission tasking. These missions entered the annals of aviation history as the longest-range combat sorties ever flown by carrier-based aircraft.

---

[5] Steve Vogel, "Gas Stations in the Sky Extend Fighters' Reach," *Washington Post*, November 1, 2001.

[6] Panel presentation on Operation Enduring Freedom by the participating carrier air wing commanders at the Tailhook Association's 2002 annual symposium, Reno, Nevada, September 6, 2002.

[7] William M. Arkin, "A Week of Air War," *Washington Post*, October 14, 2001. The JDAM's Mk 84 bomb core contains 945 lb of tritonal, which consists of solid TNT laced with aluminum for stability. The bomb's 14-in. wide steel casing expands to almost twice its normal size before the steel shears, at which point a thousand pounds of white-hot steel fragments fly out at 6,000 ft per second with an initial overpressure of several thousand psi and a fireball 8,500 deg Fahrenheit. The bomb can produce a 20-ft crater and throw off as much as 10,000 lb of dirt and rocks at supersonic speed. (David Wood, "New Workhorse of U.S. Military: A Bomb with Devastating Effects," Newhouse.com, March 13, 2003.)

[8] Robert Wall, "Targeting, Weapon Supply Encumber Air Campaign," *Aviation Week and Space Technology*, October 22, 2001, p. 28.

Two separate disciplines were consecutively employed in conducting carrier-based strike operations during Enduring Freedom. The first of these, the familiar and classic air-wing strike, was target-specific and involved complex planning and large-force tactics. It predominated during the first week and a half of the campaign. During these initial attacks, carrier air-wing weapons drops mainly featured precision munitions delivered against prebriefed fixed targets. The second discipline, time-sensitive targeting (TST) attack, was more flexible and adaptive and made for less burdensome mission planning, since the attacking aircrews would launch without prebriefed target assignments and would be given target coordinates only after getting airborne. That pattern began to predominate at around the war's eleven-day point when the Department of Defense indicated that the campaign had shifted from attacking largely fixed targets to seeking out targets of opportunity in designated engagement zones.

As the second week of Enduring Freedom gradually unfolded, numerous changes in target assignments occurred after Afghanistan was divided into engagement zones in which pop-up targets of opportunity began to emerge. In this new phase of operations, airborne forward air controllers (FAC-As) loitering overhead would identify emerging targets and then clear other aircraft to attack them upon receiving approval from either the CAOC or the airborne command and control center (ABCCC) orbiting over Afghanistan.[9] (No ground combat controllers were yet involved in the war at this point, since persistent adverse weather had prevented CENTCOM from inserting a SOF presence into Afghanistan.) The Deputy Director of Operations on the Joint Staff, Rear Admiral John Stufflebeam, explained that the engagement-zone arrangement did not precisely equate to a "free-fire, free-target environment," but rather to one in which aircraft would be directed to targets once the latter were determined to be valid. He declined to indicate how many such zones had been attacked, since that would telegraph U.S. capability. However, he said,

[9] In the F-14, both the pilot and the backseat radar intercept officer (RIO) had to be FAC-A qualified for the aircraft to perform that function.

"there isn't any part of the country that couldn't be put under an engagement zone."[10]

As strikes on emerging TSTs increasingly became the norm, orbiting attack aircraft would be routed to and talked onto those targets, sometimes including enemy vehicles moving as fast as 60 mph, by Air Force Special Operations Command (AFSOC) combat controllers on the ground. By the campaign's eleventh day, the first U.S. SOF teams had finally been introduced into Afghanistan and had linked up with various units of the indigenous Afghan Northern Alliance opposition group. Once a moving target was acquired, strike aircrews would monitor its activity until authorized to attack it. Combat controllers used laser designators to mark targets and offered reliable eyes on target to ensure that inadvertent attacks on civilians did not occur. (Most of the time, however, moving targets were designated by airborne platforms rather than by ground controllers, since the target could quickly move away from SOF personnel on the ground.)

In light of the substantially reduced AAA threat, the CAOC now cleared fighter aircraft to descend to below 15,000 ft day or night, as their pilots deemed appropriate, to attack any emerging targets that were observed to be on the move. F-14 and F/A-18 aircrews would occasionally resort to using simple free-fall bombs against approved TSTs, so long as there was no possibility of causing collateral damage, rather than spend extra time trying to determine precise Global Positioning System (GPS) target coordinates for a JDAM attack or to place a laser spot on a target with the LANTIRN pod for an LGB drop.[11] The weapon of choice against moving targets, how-

[10] Rowan Scarborough, "U.S. Splits Afghanistan into 'Engagement Zones,'" *Washington Times,* October 18, 2001.

[11] Rear Admiral Matthew Moffit, then serving in the Directorate of Air Warfare (OPNAV N78) in the Office of the Chief of Naval Operations, later noted that LGBs often outshined JDAMs in responsiveness, since having ground combat controllers available to put laser spots on targets promptly made target attack less cumbersome than having an aircrew manually enter target coordinates into the aircraft's avionics suite. ("Notes from the Precision Strike Conference, Fort Belvoir, Virginia, April 16," *Inside the Navy,* April 29, 2002.)

ever, remained the LGB, since it was all but impossible to hit such targets with unguided bombs.

Although *Kitty Hawk* was dedicated to supporting SOF operations, her ocean station was hours of flying time from Afghanistan by helicopter. Accordingly, those SOF helicopters operating off the flight deck of *Kitty Hawk* that were not configured for inflight refueling had to be refueled in Pakistan en route to their final destinations. That constraint spotlighted a need for a quick-response makeshift airfield within Afghanistan just as soon as one could be secured and made operational. Little AAA activity was observed by this point in the war because enemy gunners had learned that their positions would be bombed if they fired on the attacking aircraft. Once CENTCOM was confident that the Taliban's limited air defenses had been sufficiently degraded, the standing policy requiring that all strike aircraft have EA-6B escort jamming protection was lifted.[12]

As many as a dozen allied tankers were airborne in the war zone at any moment to support these operations. General Tommy Franks, CENTCOM's commander, asked for a fourth carrier to be deployed to the AOR to relieve *Enterprise*, which finally departed for home the first week of November, thereby ending her cruise extension that began immediately after the terrorist attacks on September 11.[13] To honor that request, the Navy found itself obliged to juggle various deployment options, with one sympathetic official commenting that "they're living with 12 carriers in a war where we need 15."[14] The Pentagon also dispatched USS *John C. Stennis* to the AOR from her home port in San Diego on November 12 to relieve *Carl Vinson*, which was operating in the Arabian Sea northeast of Masirah.[15]

---

[12] Robert Wall, "EA-6B Crews Recast Their Infowar Role," *Aviation Week and Space Technology*, November 19, 2001, p. 41.

[13] By the time *Enterprise* was relieved, 72 percent of the munitions delivered by CVW-8 had been LGBs, with 16 percent consisting of JDAMs and 12 percent laser-guided AGM-65 Maverick air-to-ground missiles.

[14] Rowan Scarborough, "Air Force Slow to Transfer Special Bomb Kits to Navy," *Washington Times*, November 7, 2001.

[15] Bill Gertz and Rowan Scarborough, "Pentagon to Send Fourth Carrier to Afghanistan," *Washington Times*, November 8, 2001.

As the campaign's endgame neared, *Theodore Roosevelt* launched scores of fighter sorties in direct support of the battle for Mazar-i-Sharif. The ensuing success on the ground made for the first tangible allied victory in Enduring Freedom, as well as a notable morale booster at a time when concerns about the campaign's halting progress had begun to mount across the board. In yet another encouraging sign of progress, there were confirmed reports, based on intercepted enemy radio traffic, that bombing in the Kandahar area had finally succeeded in killing some senior al Qaeda leaders.[16]

With the enemy's fallback redoubt in the remote Tora Bora cave complex all but obliterated by early December, some al Qaeda survivors sought to regroup once again in caves in eastern Afghanistan at Zhawar Kili and in the nearby vicinity of Khowst. That development prompted 118 consecutive attack sorties in the area over a four-day period, including strikes by numerous F-14s and F/A-18s. Some 250 bombs were dropped on caves at Zhawar Kili alone.[17] One early indication that the Pentagon was now content with the self-sufficiency of U.S. SOF units in Afghanistan was the departure of *Kitty Hawk*, which had been used as a staging base for SOF operations during the early phase of the war, for her home port in Japan the second week of December.

The first phase of bombing in Enduring Freedom ended on December 18. The week that followed was the first since the war began on October 7 in which no bombs were dropped, although numerous armed F-14s, F/A-18s, B-52s, and B-1s continued to orbit on call over Kandahar and Tora Bora to attack any possible residual al Qaeda targets that might emerge. Those aircraft were joined by Italian Navy Harriers operating off the carrier *Garibaldi* and by French Super Etendard fighters from the carrier *Charles de Gaulle*.[18] By mid-January 2002, offensive air operations over Afghanistan had largely

---

[16] James Dao, "More U.S. Troops in bin Laden Hunt; Hideouts Bombed," *New York Times*, November 19, 2001.

[17] Esther Schrader, "U.S. Keeps Pressure on al Qaeda," *Los Angeles Times*, January 8, 2002.

[18] Douglas Frantz, "Hundreds of al Qaeda Fighters Slip into Pakistan," *New York Times*, December 19, 2001.

been reduced to a trickle, and only one in ten strike sorties dropped munitions.

Two months after the rout of the Taliban and the installation of the interim successor government of Hamid Karzai, U.S. forces met their single greatest challenge of the war in an initiative that came to be known as Operation Anaconda. The Shah-i-Kot valley area in eastern Afghanistan near the Pakistani border had been under surveillance by CENTCOM ever since early January 2002, prompted by intelligence reports that Taliban and al Qaeda forces were regrouping there in an area near the town of Gardez. Over time, enemy forces continued to mass in the area, to a point where it appeared as though they might begin to pose a serious threat to the Karzai government. At that point, the U.S. Army's Combined Joint Task Force (CJTF) Mountain began planning an operation aimed at surrounding the Shah-i-Kot valley with overlapping rings of U.S. and indigenous Afghan forces, with the intent to bottle up and capture or kill the several hundred al Qaeda fighters who were thought to have congregated in the area.

Carrier Air Wing (CVW) 9 aboard *John C. Stennis*, which had taken up station in Afghan war zone in mid-December 2001, played a major part in Operation Anaconda, as did CVW-7 in *John F. Kennedy*. The sorties flown by the two air wings included some unique operations, such as F-14 formations (either two-plane sections or four-plane divisions) carrying mixed weapons loads of LGBs and JDAMs. CVW-7 F-14Bs from VF-11 and VF-143 were the first F-14s to carry and employ the JDAM and mixed ordnance loads. E-2C Hawkeyes provided airborne command and control inside Afghanistan, and EA-6Bs provided 24-hour alert jamming support for CJTF Mountain.[19] The participating air wings also, as needed, flew mixed fighter formations, with a section of F-14s and another section of F/A-18s making up a four-plane division.

---

[19] Panel presentation on Operation Enduring Freedom by the participating carrier air wing commanders at the Tailhook Association's 2002 annual symposium, Reno, Nevada, September 6, 2002.

By the end of the first week of Anaconda fighting, as allied air attacks in support of the embattled ground troops became more consistent and sustained, al Qaeda resistance tapered off and friendly forces seized control of more terrain. Carrier-based F-14s and F/A-18s contributed significantly to this support. In addition, 16 Super Etendards from the French Navy's carrier *Charles de Gaulle* took part in Anaconda by providing close air support (CAS), along with French Mirage 2000Ds operating out of Manas airfield in Kyrgyzstan.

Several air-wing commanders later recalled that Anaconda had been poorly planned from their perspective as CAS providers.[20] The operation's leaders had counted on extensive rotary-wing support, which the Army's AH-64 Apache attack helicopters proved unable to provide in the face of intense enemy fire. They also, one commander remarked, paid insufficient heed in their planning to weather considerations. The most intense early battle during the predawn hours of March 4 ended up involving a heavy combat search and rescue effort with multiple ground forward air controllers (FACs) working a very small area. The sudden and unexpected demand for air support that was occasioned by it and by ensuing battles led to an airspace congestion problem of formidable proportions, with allied aircraft frequently stacked eight miles high over the combat zone. B-52s at the highest altitude of 39,000 ft dropped JDAMs through formations of B-1 bombers and fighters orbiting at 20,000–25,000 ft, P-3s at lower altitudes, and AC-130s lower still at night, all followed by RQ-1 Predator unmanned aerial vehicles (UAVs), A-10s, and attack helicopters at the lowest operating altitudes. With multiple JDAMs repeatedly falling through this densely occupied airspace, only the tightest and most exacting air discipline, plus a considerable measure of good luck, prevented a major catastrophe.

---

[20] Panel presentation on Operation Enduring Freedom.

Carrier Air Operations in Retrospect

In all, from the start of hostilities on October 7, 2001, until the period of major combat ended in mid-March of 2002, six carrier battle groups participated in Operation Enduring Freedom (see Figure 2.1). Together, they conducted around-the-clock combat operations against enemy forces in a landlocked country more than an hour and a half's flight north of the carrier operating areas in the Arabian Sea. Those operating areas were repositioned from time to time to meet changing tactical requirements. Eventually, however, carrier-launched air missions came to average a distance of 600 nautical miles from their stations some 100–120 nautical miles south of the Pakistani coast to central Afghanistan and another 150–200 nautical miles to northern Afghanistan.

Figure 2.1
Carrier Presence on Station During Operation Enduring Freedom

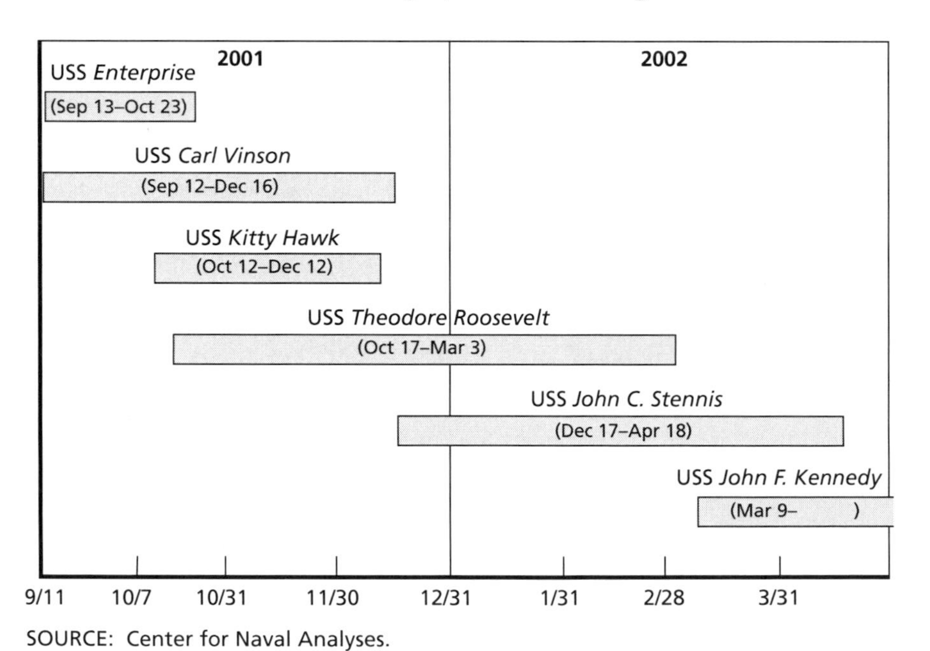

SOURCE: Center for Naval Analyses.

Throughout the war, the Navy maintained at least two carriers on station in the North Arabian Sea. When *Theodore Roosevelt* arrived on station on October 17, 2001, to relieve *Enterprise*, CENTCOM's Combined Force Air Component Commander (CFACC) had three carrier air wings available to him for nearly a week, as the presence of *Enterprise* was extended until October 23. The number of carriers deployed in the AOR did not drop to one until *John C. Stennis* departed for home on April 18, 2002, leaving *John F. Kennedy* the only carrier remaining in place to support the final wind-down of Enduring Freedom.

Enterprise remained on station in support of Enduring Freedom from September 13, 2001, until October 23, 2001. *Carl Vinson* preceded *Enterprise* in the AOR by a day, starting on September 12, 2002, and remained on station until December 16, 2001, just after the heaviest part of initial air operations had ended. *Kitty Hawk*, which deployed primarily to provide an afloat forward staging base (AFSB) for SOF helicopters and personnel, arrived on station on October 12, 2001, and remained in place until December 12, 2001. To keep her flight deck and hangar bay clear for SOF aircraft, the fighter and attack squadrons of CVW-5 remained ashore at their home station. A detachment of eight F/A-18s was kept aboard primarily to provide self-protection for the carrier, although those aircraft also eventually participated in strike operations into Afghanistan with the other deployed air wings as needed.

As noted above, *Theodore Roosevelt* arrived on station on October 17, 2001, and remained in place until Operation Anaconda commenced on March 3, 2002. *John C. Stennis* commenced operations on December 17, 2001, and remained in the AOR until April 18, 2002, nearly a month after the bombing phase of Enduring Freedom had drawn to a close. Finally, *John F. Kennedy* arrived on scene on March 9, in the midst of Anaconda's heaviest fighting, and remained as a single-carrier presence until well after the air war ended.

As the terrorist attacks were taking place on September 11, 2001, CVW-9 in *John C. Stennis* was in the midst of conducting its final predeployment workups at the Naval Strike and Air Warfare Center (NSAWC) at Naval Air Station (NAS) Fallon, Nevada. Initially slated for a January 2002 deployment, *Stennis* was promptly rescheduled to deploy on November 12 as CVW-9 accelerated its workup training at Fallon, concentrating on aircrew strike-lead certification in a now-abbreviated Air Interdiction Mission Commander course stressing classic Navy Alpha strike mission profiles and aiming for a start-of-operations date of December 17. As *Stennis* got under way heading for the AOR, CVW-9 squadron operations officers communicated daily over the SIPRNET (secure Internet protocol router network) with their counterparts already on station and engaged in combat to familiarize themselves with the daily flow pattern and prepare for what was to come. Squadron commanders and executive officers conducted similar daily SIPRNET exchanges with their engaged counterparts in the AOR. The *Stennis* battle group commander, then–Rear Admiral James Zortman, made clear to all CVW-9 principals his determination that when *Stennis* replaced *Vinson*, the transition would be so seamless that no one in the CAOC would even notice.[21]

Throughout the war, carrier-based strike assets in all participating air wings averaged around 40 actual shooter sorties a day per carrier. Fighter missions fell into the categories of preplanned strike, attacks against emerging time-critical targets, and support to ground forces, with most of the first category aggregating during the war's initial week and virtually all of the third category occurring in November 2001, during the battles leading up to the fall of Kabul, and in early March 2002 in connection with Operation Anaconda. From the campaign's second night onward, time-critical targets were also attacked, becoming the majority of target types attacked after the war's second week. By November 1, virtually all targets attacked were unbriefed time-critical targets.

After the war ended, one-third of all Navy strike sorties had been directed against interdiction targets with the remaining two-thirds providing air support to friendly ground forces. Around 80 percent of the carrier-based missions that dropped ordnance did so

---

[21] Conversation with Lieutenant Commander Nicholas Dienna, former VF-211 operations officer, at the RAND Corporation, Santa Monica, Calif., June 17, 2003.

against targets that were unknown to the aircrew before launch (see Figure 2.2). Of the Navy sorties that dropped ordnance, 84 percent were assessed as having hit at least one target, and an average of two desired mean points of impact (DMPIs) were hit by Navy sorties that dropped ordnance (see Figures 2.3 and 2.4). Of all Navy munitions dropped, 93 percent were either satellite-aided or laser-guided. Targets were attacked at all hours of the day by Navy strike fighters, with most weapon impacts occurring during the first three hours of daylight (see Figures 2.5 and 2.6).

Each carrier conducted flight operations for roughly 14–16 hours a day, with overlaps as needed to keep an average of three two-plane sections of fighters constantly over Afghanistan for on-call strikes against emerging targets. Most aircraft flew triple or quadruple deck-cycle missions, with a typical cycle duration being an hour and a

Figure 2.2
Preplanned Strikes vs. Time-Critical Targets

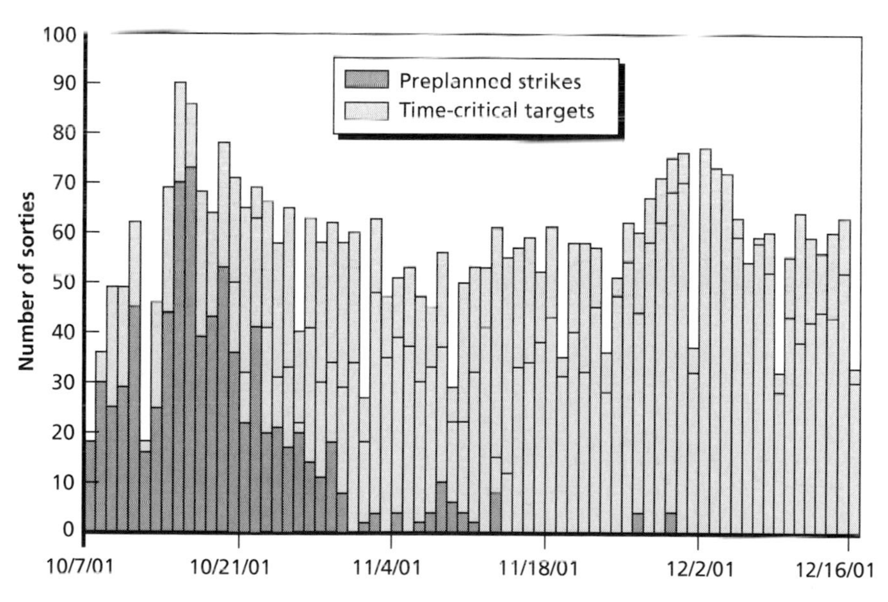

SOURCE: OPNAV N3/5 (Deep Blue).

Figure 2.3
Hit Rate of Sorties That Dropped Munitions

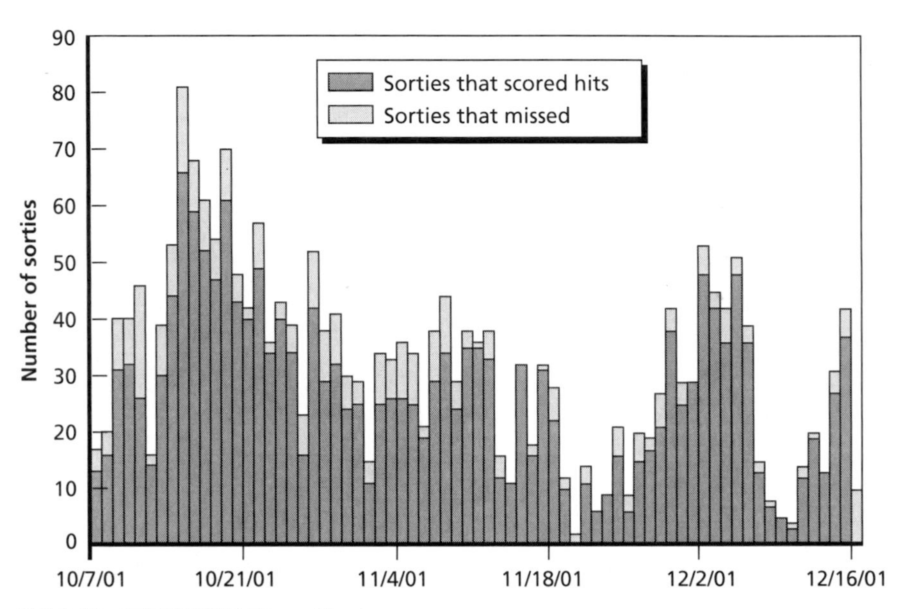

SOURCE: OPNAV N3/5 (Deep Blue).

half (or 1+30).[22] Of all missions flown, 25 percent lasted longer than five and a half hours, with some F/A-18 missions lasting up to ten hours. In a typical two-carrier operation, *Carl Vinson* served as the day carrier and *Theodore Roosevelt* performed as the night carrier. When *Vinson* departed the AOR and was replaced by *John C. Stennis*, *Theodore Roosevelt* rolled forward to become the day carrier as *Stennis*

---

[22] A 1+30 cycle is one that lasts an hour and thirty minutes from an aircraft's launch to its recovery. A 1+00 cycle lasts an hour. In the instance of a notional 1+30 cycle, while one wave of aircraft is being launched, the preceding wave that was launched an hour and a half earlier will be holding overhead, with its pilots watching their constantly dwindling fuel levels and, as may be required, conducting ecovery tanking near the carrier while the flight deck is being prepared for their recovery and as they await the signal to extend their tailhooks and commence their approach to the carrier in sequence. In this manner, 20–30 aircraft can be kept airborne at any given time and the extra space thus freed up on the flight deck can be exploited for moving (or "respotting") aircraft to prepare for the next launch.

Figure 2.4
Attacked Aim Points per Sortie

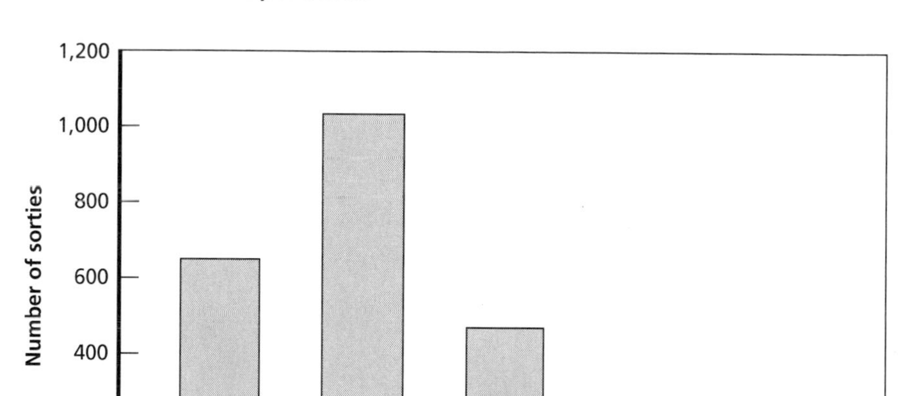

SOURCE: OPNAV N3/5 (Deep Blue)

became the new night carrier, launching and recovering sorties from 2230 to 1330 on the following afternoon. The day carrier, for her part, conducted flight operations from 1030 until 0130 on the following night. Missions were typically scheduled for two tanker hookups during the hour and a half inbound leg to the assigned CAP station over Afghanistan and two more inflight refueling plugs during the return leg.

On a typical flying day, squadron aircrews would receive CENTCOM's daily situation briefing at 0700, then receive a target-area intelligence briefing at 1000 for a noon launch. In the case of night operations, the air-wing flying schedule would be completed by 1100, at which time the aircrews would rest and then awaken for the initial mission briefing at 2130 for a scheduled midnight launch followed by an 0450 recovery. The mission briefing included those segments of CENTCOM's daily ATO for the scheduled air-wing aircraft, including call signs and all other pertinent operational

Figure 2.5
Precision-Guided vs. Free-Fall Weapons Expended

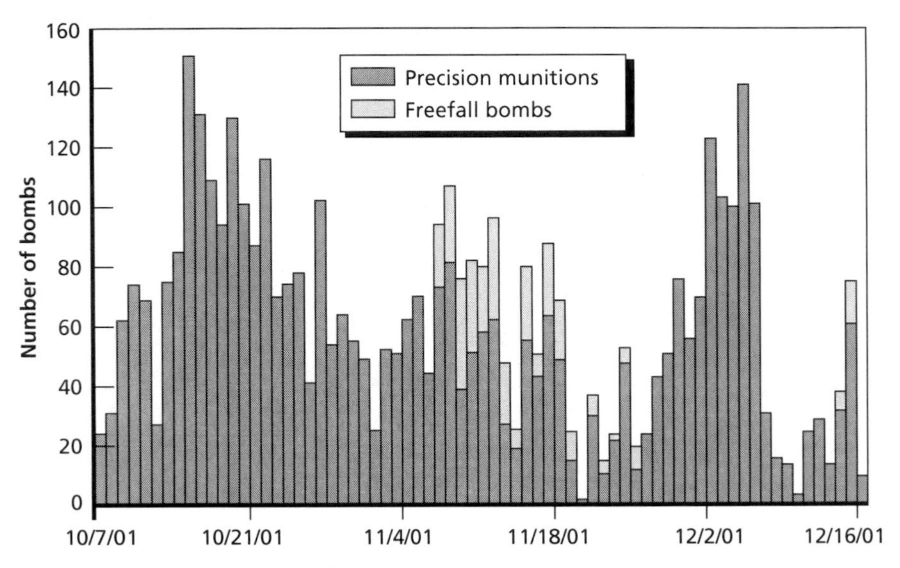

SOURCE: OPNAV N3/5 (Deep Blue).

information. The briefing included a review of rules of engagement (ROE), additional special instructions (SPINs) regarding mission execution, and safety considerations. After that review came an intelligence briefing of likely target sets that generally proved irrelevant to the targets that were actually assigned and attacked once the fighters were airborne. The operating tempo was fairly moderate, with one bow catapult and the two waist catapults typically in use. Because a deck cycle involved the launch of an outgoing strike package and the recovery of the previous launch, sortie duration had to be in multiples of 1+30 deck cycles as mission needs demanded.

Throughout Enduring Freedom, the aircrew work schedule entailed 14 days on duty followed by a day off for rest, with the normal peacetime workload being five days on and one day off. Although many Navy combat missions were of an unprecedentedly long duration, the daily combat sortie *rate* was not especially onerous. Each air

Figure 2.6
Time-of-Day Distribution of Target Attacks

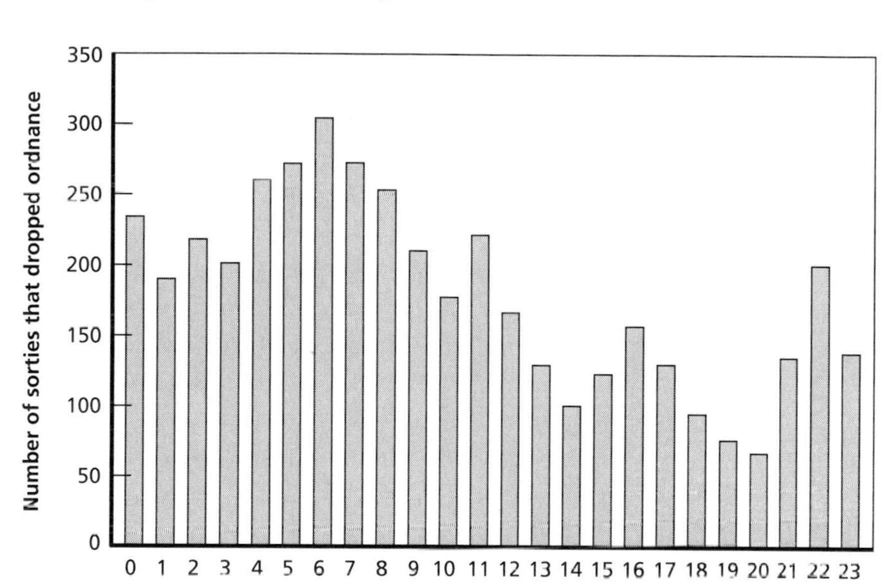

SOURCE: OPNAV N3/5 (Deep Blue).

wing flew an average of 30–40 combat sorties a day, with the maximum being around 42, since that rate was more than enough to support the CFACC's target-coverage requirements. The remaining sorties that made up the daily baseline of 90 per air wing entailed tanker, electronic warfare, command and control, and other mission support. The tyranny of distance between the carrier operating area and target area, however, made for an operating environment that was extremely unforgiving. Aircrews would occasionally find themselves in a "tank or die" situation, with tanker hook-ups occurring after an aircraft's fuel state had fallen so low that the aircraft would not have been able to make it to Pakistan in the event that a tanker had been unavailable. (A standing rule proscribed any U.S. fighter from landing in Pakistan unless its fuel state was such that the aircraft would flame out otherwise.) Because of the long-duration combat sorties, fighter aircrews averaged 70 flight hours a month, well over twice the normal peacetime rate.

Key Operational Achievements

As indicated by statistics compiled by the CAOC during the 76 days of bombing between October 7, when Enduring Freedom began, and December 23, when the first phase of the war ended after the collapse of the Taliban, some 6,500 strike sorties were flown by CENTCOM forces altogether, out of which approximately 17,500 munitions were dropped on more than 120 fixed targets, 400 vehicles and artillery pieces, and a profusion of concentrations of Taliban and al Qaeda combatants. Of the total number of allied munitions expended, 57 percent were precision-guided munitions. U.S. carrier-based strike fighters accounted for 4,900 of the strike sorties flown during that period, making up 75 percent of the total (see Figure 2.7). More than half of those sorties were flown by Navy and Marine Corps F/A-18s (see Figure 2.8).

Altogether through the spring of 2002, Navy and Marine Corps tactical aircraft flew more than 12,000 combat sorties (approximately 72 percent of all flown in Enduring Freedom), with maritime forces accounting for more than half of all precision munitions expended. Because of the need for an extended deployment due to newly discovered readiness problems in *John F. Kennedy* deployed elsewhere, *Theodore Roosevelt* broke a record after spending 153 days at sea without a port call, having conducted more than five months of nearly constant strike operations. Her embarked CVW-1 flew more than 10,000 sorties, logged more than 30,000 flight hours, and dropped some 1.7 million pounds of ordnance.[23]

---

[23] Scott C. Truver, "The U.S. Navy in Review," *Proceedings*, May 2003, p. 91. The longest post–World War II deployment record was set by USS *Coral Sea* (CV-43), which logged 329 days at sea during a 1964–1965 Western Pacific deployment.

Figure 2.7
Strike Sorties Through December 2001 by Service

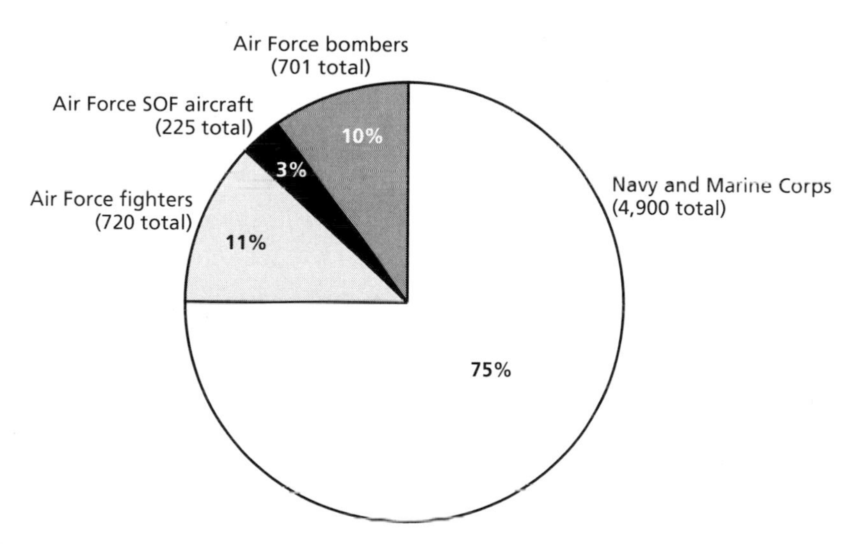

SOURCE: *Sea Power*, March 2002.

As many in the naval aviation community were among the first to acknowledge, without nonorganic Air Force and RAF tankers to provide inflight refueling support, the Navy's carrier air wings could not have participated in Operation Enduring Freedom beyond the southernmost target areas in Afghanistan.[24] Yet that admitted fact of joint and combined warfare life is neither here nor there. What matters above all else is that carrier-based air power, more than at any other time in the post-cold-war era since Desert Storm, showed beyond doubt its indispensability as a vital component of U.S. force projection capability owing to the remoteness of the Afghan theater—even if the Navy's carrier air wings, like the Air Force's complementary air assets, could not have swung the outcome by themselves.

---

[24]This point was made emphatically by then–Rear Admiral James Zortman, USN, who commanded the USS *John C. Stennis* battle group, during a panel discussion on carrier operations in Enduring Freedom at the Tailhook Association's 2002 annual symposium, Reno, Nevada, September 6, 2002.

Figure 2.8
Strike Stories Through December 2001 by Aircraft Type

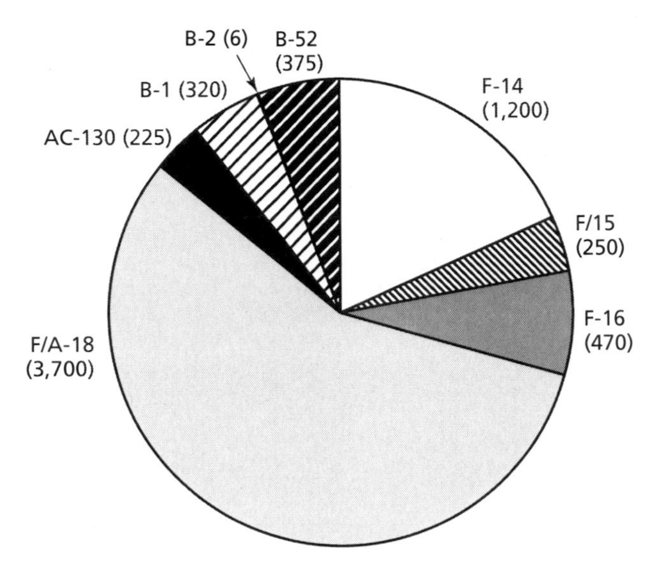

SOURCE: *Sea Power*, March 2002.

The carrier air wings committed to Operation Enduring Freedom sustained a substantial, if far from record-breaking, sortie rate throughout the campaign. Each flew a baseline of 90 sorties a day, and each met the CFACC's tasking every day. Their commanders uniformly felt that this sortie rate was maintainable for an extended length of time. To cite but one case in point, VFA-97, flying the oldest F/A-18s in the Navy's operational inventory, had pilots averaging 72 flight hours a month, as compared to 30 a month during a normal peacetime deployment. The air-wing squadrons typically sustained utilization rates of more than two and a half times their normal programmed flying hours.

Fully mission-capable (FMC) rates were likewise better during Enduring Freedom than they had been before the war started. From the commencement of the campaign on October 7 through December 16, 2001, when *Carl Vinson* was relieved by *John C. Stennis* and the first phase of the war was over, the Navy's Task Force 50 had flown nearly 4,000 strike sorties in Enduring Freedom altogether. As impressive as that number may appear to be in the abstract, it should be noted that the participating carriers and air wings could have generated considerably more had additional sea-based sorties been required to meet the CFACC's target-coverage needs. More important from an operational perspective, however, the Afghan experience affirmed that the emerging metric that now matters most in modern air warfare, thanks to the revolution in precision weapons that can be delivered from safe standoff ranges, is not sortie generation but actual ordnance placed on targets.

Operation Enduring Freedom also featured the interplay of a veritable constellation of intelligence, surveillance, and reconnaissance (ISR) systems that proved pivotal in enabling the joint air campaign. In a typical ISR fusion scenario involving carrier air assets, an electronic intelligence (ELINT) aircraft orbiting near an enemy target area of interest would note a spike in communications traffic coming from a known Taliban location. An RQ-1 Predator UAV would then be sent to the vicinity for a closer look, streaming real-time video of the building to targeteers in the CAOC in Saudi Arabia and at CENTCOM headquarters in Tampa, Florida. An F-14 pilot and RIO orbiting overhead nearby would search for additional signs of activity and would visually confirm the suspected target to be a valid one. A staff operator in the CAOC, via the E-3 airborne warning and control system (AWACS) aircraft, would then read exact target coordinates to the F-14's RIO, who would finally program a JDAM for the assigned DMPI on the building as soon as approval to drop was received.

At the war's 45-day point, the Joint Staff's spokesman, Rear Admiral Stufflebeam, reported that around 10,000 munitions of all types had been employed, a rate roughly comparable to that of the earlier Operation Allied Force over Serbia and Kosovo in 1999, which saw 23,614 munitions expended over its 78-day course. A major difference between the two campaigns, however, was in the overall percentage of precision weapons employed (close to 60 percent in Enduring Freedom, as compared to only 35 percent in Allied Force).[25] As noted above, of all the participating carrier air wings, 84 percent of the weapons dropped by them hit their designated aim points and 93 percent of those weapons were precision-guided. On average, two DMPIs per aircraft were hit on each sortie.[26] This represented a quantum improvement over the Navy's performance in Desert Storm, when only the A-6E had an autonomous precision-attack capability. In one well-known instance, an AFSOC combat controller attached to an Army Special Forces A-Team (call sign Texas 17) targeted a bridge three miles southeast of Kandahar as Taliban members huddled underneath it for protection, confident in their knowledge that bridges were not approved targets. Thanks to this timely SOF cueing, an F/A-18 pilot was able to skip an AGM-65 laser-guided Maverick missile under the bridge, killing the enemy combatants who had sought refuge there while leaving the bridge itself undamaged.[27]

Navy F-14s transmitted and received imagery from allied SOF units on the ground using the aircraft's Fast Tactical Imagery (FTI) system. That system coupled the fighters for the first time with SOF teams deployed with Northern Alliance forces around Kabul. The SOF teams communicated with F-14 crews via laptop computers. First used during Operation Southern Watch over Iraq in early 1999 and again during Operation Allied Force later that same year, FTI provided a day and night standoff transmission system enabling F-14 crews to send and receive imagery to and from SOF units, thus giving the aircraft a near-real-time two-way imagery capability. Imagery from the ground could be sent back to the carrier for either clearance

---

[25] Tony Capaccio, "Sixty Percent of Bombs Dropped on Afghanistan Precision-Guided," Bloomberg.com, November 20, 2001.

[26] Vice Admiral John B. Nathman, USN, "We Were Great: Navy Air in Afghanistan," *Proceedings*, U.S. Naval Institute, March 2002, pp. 94–96.

[27] Vernon Loeb, "Afghan War Is a Lab for U.S. Innovation," *Washington Post*, March 26, 2002.

to attack, if needed, or for almost instant poststrike battle-damage assessment (BDA).[28]

An arrangement was worked out during Enduring Freedom whereby allied SOF units were also able to provide fine-resolution target imagery to the F-14s. The FTI system uses a two-hour tape recorder and transceiver enabling the RIO to capture and store any video imagery acquired by the aircraft's systems, including the TARPS pod, the pilot's head-up display (HUD), the Television Camera System (TCS, essentially a daytime zoom lens), and the LANTIRN pod. Enhancements made to the system allowed the F-14 to decrease the imagery transmittal time from 90 seconds to around 25–30 seconds, with an average data cycle time, including storage and transmission to a carrier, of around two minutes. Such use of the F-14's FTI by U.S. SOF teams on the ground allowed for a substantial shortening of what was informally called the kill chain.[29]

The F-14 also, for the first time during Enduring Freedom, showed its ability to derive and transmit accurate coordinates acquired by its LANTIRN targeting pod to an inbound B-52, whose crew, in turn, would use the information to drop JDAMs with great accuracy on approved targets. Using a feature of the LANTIRN pod called T3, for Tomcat Tactical Targeting, the RIO could determine a target's exact geographic coordinates by that technique and then pass those coordinates on to Air Force F-16s that lacked GPS receivers or LANTIRN pods and enable their pilots to drop GPS-guided cluster bomb units (CBUs). Aircrews who used it said that the technique worked consistently well.

---

[28] Robert Wall, "F-14s Add Missions in Anti-Taliban Effort," *Aviation Week and Space Technology*, November 19, 2001, p. 38.

[29] Frank Wolfe, "Navy F-14s Able to Transmit, Receive Imagery from Green Berets in Afghanistan," *Defense Daily*, August 1, 2002, p. 1. All of the Navy's remaining F-14s are equipped with the capability. The F/A-18C currently lacks it, but the F/A-18E/F Super Hornet began receiving it in early 2003.

Lessons of the Afghan Air War

By September 11, 2002, nearly a year after the commencement of Operation Enduring Freedom, eight of the Navy's 12 carriers had participated in strike operations over Afghanistan against the Taliban and al Qaeda. This sustained contribution of naval aviation showed the ability of as many as four carriers at a time to maintain a sufficient sortie rate to enable a constant armed airborne presence over a landlocked theater located more than 400 nautical miles away from the carriers' operating stations in the North Arabian Sea. In so doing, it roundly disconfirmed suggestions voiced by some critics only a few years before that the Navy's carrier force lacked the capability to turn in such a performance.[30]

In conducting combat operations throughout the five-month course of major fighting in Enduring Freedom, the participating air wings showed the substantially improved capability that naval strike aviation had acquired since the 1991 Persian Gulf War. The predominant use of precision munitions made the Afghan air war the most precise naval bombing effort in history up to that time, with LGBs delivered by Navy strike aircraft consistently hitting exactly where they were directed to hit. In this regard, the F-14 with its LANTIRN targeting pod proved all but indispensable. The war also yet again showcased naval aviation's recently acquired FAC-A capability that the F-14 first demonstrated during Operation Allied Force in 1999, along with the heightened operational value of the F-14-TARPS combination as the last remaining U.S. manned high-speed reconnaissance asset. In Enduring Freedom, unlike the earlier case of Desert Storm, the carriers could receive the daily ATO electronically, and their air wings conducted precision-strike operations almost exclusively. Those operations featured digital communications, network-centric execution, and a greatly increased integration of carriers into the digital data stream.

---

[30] See, for example, Rebecca Grant, "The Carrier Myth," *Air Force Magazine*, March 1999, p. 26.

Furthermore, thanks to the much-expanded bandwidth now available to each carrier battle group, as well as to the substantial progress that had been made in interservice cooperation since the 1991 Gulf War, naval aviation in general and the Navy's carrier air wings in particular were thoroughly integrated into CFACC planning, fully represented in the CAOC—all the way up to the position of deputy CFACC, and employed for the first time to their fullest potential in the CFACC's daily ATO tasking. Training efforts over the preceding decade had paid off, especially in the precision-strike and time-sensitive targeting arenas. Then–Rear Admiral Zortman, who served as the *John C. Stennis* battle group commander during that carrier's involvement in the war from mid-November 2001 through Operation Anaconda the following March, later observed that the consistently effective performance of the Navy's air wings in Enduring Freedom did not "just happen" but rather was a telling testament to the quality of their equipment and conditioning.[31]

Perhaps most important of all, the close meshing of carrier- and land-based air involvement in Enduring Freedom, as well as the unprecedentedly prominent role played by the Navy in the CAOC throughout the war, bore witness to a remarkable transformation that had taken place during the preceding years since Desert Storm by way of a gradual and largely situation-driven convergence of Navy and Air Force thinking with respect to force employment at all levels of war. In previous years, "jointness" typically meant little more than the concurrent and usually uncoordinated participation of two or more services in a military operation. Yet as early as 1994, motivated by the post-cold-war Navy needs identified by the 1991 Gulf War, then–Vice Admiral Owens, at the time Deputy CNO for Resources, Requirements and Assessments (OPNAV N8), introduced a new approach to Navy force planning aimed at increasing the service's combat leverage by seeking synergistic involvement with the Air Force, in particular, when it came to expanded battlespace awareness, the military exploitation of space, and making the most of centralized com-

[31] Comments at the Tailhook Association 2002 annual symposium, Reno, Nevada, September 6, 2002.

mand in joint air operations. Two commentators on that history turned out to have been more than a little prescient when they predicted, on the very eve of the September 11 terrorist attacks, that the coming year would witness "a triumph of the synergistic view of jointness . . . where the Navy and Air Force are concerned," with the result being the "closing of a promise-reality gap" that would yield "effects-based capabilities that are good for our regional commanders in chief and right for our nation."[32] That prediction was more than amply borne out by the Enduring Freedom experience that ensued starting only a few weeks thereafter.

There is no question that without the contribution of carrier-based air power, the success of Enduring Freedom would have taken longer to achieve—perhaps substantially longer. It is a misstatement, however, to suggest that the Afghan war represented the first instance in which carrier-based air power "created conditions for subsequent success in the employment of land-based tactical aircraft."[33] What enabled the employment of land-based theater air power in Operation Enduring Freedom was nothing more or less than the eventual securing by the U.S. government of the needed forward basing approvals for the latter to operate in the region. (Even when they did, however, only relatively few sorties were flown from those bases because of the great distances to target involved.) Until that occurred, land-based Air Force heavy bombers operated concurrently with carrier aviation and, as noted above, dropped the vast majority of precision JDAMs. A more compelling conclusion in this vein, as then–Vice Admiral John Nathman was quick to point out, was how Enduring Freedom showed that carrier air power is now immediately employable, commands sovereignty, and can enable and assist joint-force operations across the board.[34]

---

[32] Major General John L. Barry, USAF, and James Blaker, "After the Storm: The Growing Convergence of the Air Force and Navy," *Naval War College Review*, Autumn 2001, p. 130.

[33] Milan Vego, "What Can We Learn from Enduring Freedom?" *Proceedings*, July 2002, p. 30.

[34] Nathman, "We Were Great."

That fact was punctuated not only by the remarkable results that were achieved by pairing carrier-based precision attack assets with the enabling power of U.S. SOF teams on the ground but also by the dedicated support that was provided to those SOF teams by USS *Kitty Hawk*. Indeed, the use of *Kitty Hawk* as a staging base for SOF helicopters exemplified the versatility of aircraft carriers as force-projection platforms by showcasing their potential for performing missions other than launching fixed-wing aircraft.[35]

In sum, the Navy's contribution to Enduring Freedom proved the synergy of carrier-based air power with land-based Air Force bombers and tankers and forward-based Air Force fighters, with the enabling ability of friendly SOF personnel on the ground, and with the myriad other combat-support assets of all services, both sea-based and land-based, in producing the desired joint- and combined-force outcome in the shortest time and with the least loss of friendly and noncombatant enemy life. That contribution was dominated by around-the-clock precision-strike operations at a constant high level of intensity, by relentless time-sensitive attacks against emerging targets of opportunity, and by effective work with high-technology terminal attack controllers and agile SOF combatants on the ground. In each of these respects and more, the performance of U.S. carrier-based strike aviation was not just an impressive success story in its own right. It also turned out to have been a timely dress rehearsal for the more intense and demanding operations that the air wings embarked in USS *Kitty Hawk* (CV-63), *Constellation* (CV-64), *Nimitz*

[35] A precursor to this use of *Kitty Hawk* as an AFSB in Enduring Freedom was an experiment involving a loadout change in early 1993 when *Theodore Roosevelt* removed one F-14 squadron and her air wing's S-3 squadron and embarked in their place a Marine Air-Ground Task Force (MAGTF) that practiced such missions as a noncombatant evacuation operation (NEO) and an air-assault raid. (Commander J. D. Oliver, "Use the Carriers or Lose Them," *Proceedings*, September 1993, p. 70.) Another occurred in 1994 during Operation Restore Democracy in Haiti, when USS *America* (CV 66) and USS *Dwight D. Eisenhower* (CVN-69) were temporarily stripped of their air wings to transport the Army's 160th Special Operations Aviation Regiment and the aviation brigade of the Army's 10th Mountain Division, respectively.

(CVN-68), *Theodore Roosevelt* (CVN-70), *Abraham Lincoln* (CVN-72), and *Harry S. Truman* (CVN-75) would soon encounter in Operation Iraqi Freedom.

CHAPTER THREE

Operation Iraqi Freedom

If Operation Enduring Freedom had been tailor-made for deep-attack carrier air operations, the three-week period of major combat in Operation Iraqi Freedom that ensued a year later against Saddam Hussein was no less so, at least with respect to missions launched into Iraq from operating areas in the eastern Mediterranean. Unlike the case of Operation Desert Storm more than a decade before, both Saudi Arabia and Turkey refused to allow the United States and its coalition partners the use of their bases for conducting offensive operations. Their refusal created an access problem at the eleventh hour that the nation's carrier air assets were ideally suited to address. Moreover, the Navy had just successfully demonstrated an unprecedented carrier surge capability during the six-month air war over Afghanistan. Finally, unlike the cases of both Desert Storm and Enduring Freedom, it now had a full panoply of precision munitions, including satellite-aided JDAMs that were indifferent to weather.

By the end of the first week of March 2003, the Navy had two carriers, USS *Theodore Roosevelt* and *Harry S. Truman*, on station in the eastern Mediterranean and three more—*Kitty Hawk*, *Constellation*, and *Abraham Lincoln*—deployed in the Persian Gulf along with their embarked air wings, each of which included around 50 strike aircraft.[1] In addition, USS *Nimitz* was en route to the Persian Gulf to

[1] Robert Burns, "U.S. Gulf Force Nears 300,000 as Commander, Bush Consult," *Philadelphia Inquirer*, March 5, 2003.

relieve *Abraham Lincoln*, which had already been on deployment for an unprecedented nine months.

Once it became clear that Turkey would not allow the use of its territory and airspace for supporting allied combat operations, CENTCOM began moving some ships from the eastern Mediterranean to the Red Sea on March 16 so that they would be positioned to fire Tomahawk land attack missiles (TLAMs) at Iraq through Saudi airspace when the war commenced.[2] *Theodore Roosevelt* and *Harry S. Truman* also redeployed to a new location in the eastern Mediterranean, from which they could use alternative routing for providing around-the-clock strike-fighter coverage for the SOF troops who would soon be engaged in combat in western Iraq.

On March 17, President Bush gave Saddam Hussein and his two sons 48 hours to leave Iraq and promised an attack "commenced at a time of our choosing [with the] full force and might" of the assembled coalition if the Iraqi ruler failed to comply.[3] By that time, the president had given up on any further attempt to get a second UN Security Council resolution endorsing the impending war, since it was all but certain that however many member states would have backed it in the final showdown, France would have made good on its threat to veto it.[4] As the start of combat operations neared, the commander of the *Constellation* battle group, Rear Admiral Barry Costello, indicated that planning to coordinate U.S. and British air attacks had been completed: "The bottom line is we are finished planning. We're ready today if required to execute the mission."[5]

---

[2] Peter Baker, "Marine Predicts Brief Bombing, Then Land Assault," *Washington Post*, March 17, 2003.

[3] Dana Milbank and Mike Allen, "President Tells Hussein to Leave Iraq Within 48 Hours or Face Invasion," *Washington Post*, March 18, 2003.

[4] The extensive diplomatic posturing and back-and-forthing that preceded the Bush administration's final decision to go to war is well covered in John Keegan, *The Iraq War*, New York: Alfred A. Knopf, 2004, pp. 88–125

[5] David Lynch and John Diamond, "U.S., British Forces Are 'Ready Today' for Invasion," *USA Today*, March 17, 2003.

The initial concept was for the air and ground offensives to kick off more or less concurrently, with a view toward overwhelming the central nervous system of Iraq's highly centralized political and military establishment.[6] In this concept, a heavy round of air attacks would be followed in close succession by allied ground forces advancing in strength to secure such time-sensitive objectives as the oil fields in southern Iraq. On the night of March 19, however, this plan was preempted by a last-minute attempted decapitation attack launched several hours after the intelligence community had just learned of a "high probability" that Hussein would be closeted with his advisers for several hours in a private residence at a location in a part of southern Baghdad known as Dora Farms.[7] CENTCOM's master war plan did not envisage this attack. On the contrary, the attack was a rapidly improvised response to a pop-up target of opportunity—and one that was thought to offer the possibility of being outcome-determining from the very start were it to be based on good intelligence and prove successful.

The attack against the three-building compound was conducted shortly before sunrise by 40 satellite-aided TLAMs launched from Navy ships and by two Air Force F-117 stealth attack aircraft, each of which dropped two EGBU-27 2,000 lb LGBs on the target. Three Navy EA-6B Prowlers supported the F-117s by jamming enemy integrated air defense system (IADS) radars.[8] The attack, ultimately unsuccessful because of misinformed intelligence, was limited in scope and was definitely "not the start of the air campaign," in the words of

[6] For the most detailed and authoritative coverage of the campaign planning for Operation Iraqi Freedom and its many iterations before the war finally commenced, see Bob Woodward, *Plan of Attack*, New York: Simon and Schuster, 2004; and General Tommy Franks with Malcolm McConnell, *American Soldier*, New York: Regan Books, 2004.

[7] Barton Gellman and Dana Priest, "CIA Had Fix on Hussein," *Washington Post*, March 20, 2003.

[8] Todd S. Purdum, *A Time of Our Choosing: America's War in Iraq*, New York: Times Books, 2003, p. 110; and Richard R. Burgess, "Sea Services Strike for Iraqi Freedom," *Sea Power*, April 2003, p. 32.

a senior U.S. military official.[9] A day later, during the night of March 20, allied strike and combat-support aircraft were launched both from carriers and from land bases to initiate the air war ever so tentatively. The first two days of strikes were directed mainly against Republican Guard headquarters facilities and against other targets, with the intended goal of trying to separate the Iraqi people from the regime.

The actual commencement of preplanned offensive air operations, called A-Hour, occurred on the night of March 21 with the beginning of large-scale air attacks against Iraq that totaled more than 1,700 sorties, including 700 strike sorties against roughly 1,000 aim points. During the first two days of the air war, aircraft from the two carriers in the eastern Mediterranean were routed over Egypt and Saudi Arabia instead of over Turkey. Finally, on March 23, Turkey relented and gave the allies access to its airspace, which made for more direct aircraft routing. This long-sought approval granted for the use of Turkish airspace allowed the carrier-based strike aircraft operating out of the eastern Mediterranean to use a transit corridor over Turkey and thus avoid a more circuitous route. Thanks to that, the strikers could refuel over Turkey and northern Iraq.[10]

During the war's initial days, allied combat aircraft flew a total of between 2,100 and 2,300 sorties a day. For all the early media talk about an impending air campaign that would be marked by "shock and awe," coalition air strikes were actually quite measured as CENTCOM removed hundreds of planned targets from its initial list in an effort to limit noncombatant casualties and unintended damage to infrastructure.[11] The main goal of those early allied air attacks was to disable air defense facilities around Baghdad, Basra, and Mosul and to destroy key command and control nodes. Limited air strikes were

---

[9] David E. Sanger with John F. Burns, "Bush Orders an Assault and Says America Will Disarm Foe," *New York Times*, March 20, 2003.

[10] Philip P. Pan, "Turkish Leader Makes Request on Airspace," *Washington Post*, March 20, 2003. See also Richard Boudreaux, "Two Errant Missiles Fall in Turkey," *Los Angeles Times*, March 24, 2003.

[11] Michael R. Gordon, "The Goal Is Baghdad, But at What Cost?" *New York Times*, March 25, 2003.

also conducted against Iraqi artillery and surface-to-surface missile positions in southern Iraq.[12] In support of those objectives, the five participating carrier air wings operated around the clock, with *Theodore Roosevelt* taking the night shift in the Mediterranean and *Constellation* taking the night shift in the North Persian Gulf.

The Air War Unfolds

After several days of preplanned air attacks against government buildings and other fixed targets, allied air strikes shifted to engaging Iraqi fielded forces, including those of the six Iraqi Republican Guard divisions that were deployed in and around Baghdad. Air attacks now focused on "tank plinking" in designated kill boxes. This shift in emphasis presented a major targeting challenge, since the Republican Guard divisions, having learned from the Serb experience two years earlier in Operation Allied Force, did not array their tanks in battle formation but rather dispersed them under trees and in the farming villages of the Euphrates River valley.[13]

By the end of the war's sixth day on March 25, a heavy sandstorm slowed the pace of allied ground operations considerably. The sandstorm also severely affected carrier air operations in the Persian Gulf. Aboard *Abraham Lincoln*, desert dust penetrated aircraft inlets and orifices, caused damage to canopies and engines, and occasioned some harrowing aircraft recoveries. Carrier-landing approaches into whiteouts occurred more than a few times, with cockpit videotapes showing eye-watering arrested landings that were performed flawlessly in less than a half-mile visibility. Landing signals officers (LSOs) would talk each aircraft down the centerline of the carrier's recovery area, sometimes with the aircraft becoming visible to the LSO only seconds before it trapped. Air-wing tanker squadrons flew twice their

---

[12] Greg Jaffe, "Plan Is to Cut Off Top Officers While Allies Strike Air Defenses," *Wall Street Journal*, March 20, 2003.

[13] Thomas E. Ricks, "Unfolding Battle Will Determine Length of War," *Washington Post*, March 25, 2003.

normal number of sorties to refuel fighters that were orbiting overhead waiting for openings through which they might penetrate and recover aboard ship.[14] These recovery tankers transferred fuel in turbulence at altitudes as high as 30,000 ft, where inflight refueling operations would not normally take place. In all, six launch and recovery cycles on the three carriers in the Persian Gulf had to be cancelled because of persistent airborne grit, lightning, and wind shear.

The *shamal*, as it was called in Arabic, continued for three days, with sustained winds of 25 knots gusting to 50 and visibility often less than 300 ft. The Navy did not dramatically reduce its overall sortie rate during the sandstorm, however, and the coalition continued to fly as many as 2,000 sorties a day. As columns of Republican Guard vehicles attempted to move under what their commanders wrongly presumed would be the protective cover of the *shamal*, allied air strikes destroyed a convoy of several hundred Iraqi vehicles that were believed to be ferrying troops of the Medina Division toward forward elements of the U.S. Army's 3rd Infantry Division encamped near Karbala, about 50 miles south of Baghdad. Air Force and Navy aircrews used satellite-aided JDAMs for these attacks, since LGBs would not guide in those prohibitive weather conditions.[15]

Once the *shamal* abated, allied air attacks resumed their initial intensity. As the attacks were ramped up again, more than half (480 out of 800 on one day) were directed against Republican Guard units.[16] Air strikes intensified even more at the end of March. This time, they focused not only on fielded Iraqi ground forces but also on the telephone exchange, television and radio transmitters, and government media offices.[17] Every effort was made to avoid the destruc-

[14] Carol J. Williams, "Navy Does Battle with Sandstorms on the Sea," *Los Angeles Times*, March 27, 2003.

[15] Peter Baker and Rajiv Chandrasekaran, "Republican Guard Units Move South from Baghdad, Hit by U.S. Forces," *Washington Post*, March 27, 2003.

[16] Dave Moniz and John Diamond, "Attack on Guard May Be Days Away," *USA Today*, March 31, 2003.

[17] Anthony Shadid, "In Shift, War Targets Communications Facilities," *Washington Post*, April 1, 2003.

tion of infrastructure that would be essential for postwar reconstruction.

The intense Iraqi IADS activity that had been anticipated in response to allied opening-night air operations never materialized. On the contrary, Iraqi acquisition and tracking radars did not emit even once over the course of nearly a month of virtually nonstop air attacks, and allied aircraft encountered almost no SAM or AAA fire from Iraq's now thoroughly intimidated air defenses. The SAM and AAA fire that did occur was invariably unguided, although these barrages came increasingly close to the targeted aircraft on some occasions. Most of this sporadic antiaircraft fire, however, occurred after allied bombs had already hit their targets.[18]

On April 6, CENTCOM declared that allied air supremacy had been achieved over all of Iraq. Three days later, the CFACC, Air Force Lieutenant General T. Michael Moseley, reported that the CAOC was running out of worthwhile targets.[19] Saddam Hussein's regime finally collapsed on April 9 as U.S. forces drove through the streets of Baghdad, encountering only scattered resistance as thousands of residents poured into the streets to celebrate the regime's defeat. By mid April, allied combat and combat-support sorties were down to around 700 a day, only a third of the peak rate that had been sustained throughout most of the three-week war. The *Kitty Hawk* and *Constellation* carrier battle groups and their embarked air wings were sent home from the Persian Gulf on the war's 27th day, leaving one carrier remaining in the Gulf and two in the eastern Mediterranean within range of Iraq should their services be required by CENTCOM.

In the end, it was the combination of allied air power and the speed of the allied ground advance that got coalition forces to Baghdad before the enemy could establish an adequate defense. Looking

---

[18] David A. Fulghum, "New Bag of Tricks: As Stealth Aircraft and Northern Watch Units Head Home, Details of the Coalition's Use of Air Power Are Revealed," *Aviation Week and Space Technology*, April 21, 2003, p. 22.

[19] Carla Anne Robbins, Greg Jaffe, and Dan Morse, "U.S. Aims at Psychological Front, Hoping Show of Force Ends War," *Wall Street Journal*, April 7, 2003.

back over the experience, the deputy CFACC, then–Rear Admiral David Nichols, later remarked that all the way through what he called Phase III of Operation Iraqi Freedom (Phase IV was the subsequent counterinsurgency and stabilization operation that continues to this day), "we were much more successful than even the most optimistic among us had predicted. We moved farther and faster than projected, and our combined-arms fires set new standards for persistence, volume, and lethality, day and night in all-weather conditions. The Iraqi military tried but could not react to the tempo we set on the battlefield. By the time they made a decision to do something, we had foreclosed that option."[20]

Tanker Troubles

Meeting the CFACC's inflight refueling needs proved to be a greater challenge throughout the first three weeks of Iraqi Freedom than it had been during the earlier Desert Storm campaign as the shortage of available tankers seriously complicated both planning and execution. This challenge was further exacerbated by the need to distribute the 200 available U.S. Air Force and RAF tankers over 15 bases that often had severe limitations on pipelines, in contrast to the Desert Storm precedent, when 350 tankers operated out of only five bases and with unlimited direct pipeline fuel available to coalition forces.

Only the three shore bases that had been made available to allied strike aircraft in Kuwait were situated close enough to allow land-based strike fighters to reach deep into Iraqi airspace without refueling. All of the other bases that were within reasonable striking range of Baghdad required inflight refueling of attacking aircraft during both ingress and egress. With hundreds of allied strike aircraft airborne at any given moment, there were often times when queues of fighters behind tankers became so long that some pilots had to abort their missions because of insufficient fuel remaining to continue

[20] Rear Admiral David C. Nichols, Jr., USN, "Reflections on Iraqi Freedom," *The Hook*, Fall 2003, p. 3.

holding for a scheduled tanker connection. A related concern entailed positioning tankers far enough forward and timing tanker connections in a way best assured of preventing strike aircraft from lining up in a queue and thus becoming inviting targets for enemy air defenses.

Another tanker-related problem had to do with fuel movement on the ground at those tanker beddown sites where inadequate fuel lines required that jet fuel be trucked in from the outside. For strike operations into northern Iraq from the two carriers in the eastern Mediterranean, this dependence on local fuel trucks increased tanker turnaround times and reduced tanker availability. As the director of operations in the CAOC, Navy Captain David Rogers, observed on this point, "with 15 beddown sites for 200 tankers, it's a more complex shell game than in Desert Storm."[21]

Still another problem stemmed from the Turkish government's denial of approval for coalition aircraft to operate out of Turkey to conduct combat operations, which caused CENTCOM to lose about 25 percent of its originally planned bases from which to stage tanker missions. The allied combat aircraft most heavily affected by this eleventh-hour denial of access to Turkey were those from the two air wings that were deployed in the eastern Mediterranean. Those aircraft were forced to count on initially receiving little nonorganic tanker support. In addition, with more refueling taking place closer to the war zone, the placement of tanker orbits necessarily became more complicated. In view of the continuing enemy IADS threat, it was not feasible to maintain predictable tanker tracks each day with large stacks of aircraft waiting to take on fuel. That required more tanking stations and often occasioned greater difficulty for strike aircrews in connecting with their assigned tanker. Furthermore, a tanker pilot might break off a refueling if the aircraft was in imminent danger of being fired upon.

The two carriers that had been on station in the Persian Gulf for the longest time, *Abraham Lincoln* and *Constellation*, were selected to

---

[21] David A. Fulghum, "Tanker Puzzle: Aggressive Tactics, Shrinking Tanker Force Challenge Both Planners and Aircrews," *Aviation Week and Space Technology*, April 14, 2003, p. 26.

receive the greatest possible Air Force tanking support. In contrast, the air wing embarked in the last carrier to arrive on station, *Kitty Hawk*, was given a predominant CAS role in southern Iraq in support of the U.S. Army's V Corps and the 1st Marine Expeditionary Force and was tasked to fulfill that role autonomously. That meant that *Kitty Hawk*'s organic tanking capability would be burdened to the fullest. For the first time since Operation Desert Storm, strike aircraft in CVW-5 were allowed to return to ship with minimum fuel, occasioning an employment plan that leveraged the advantages of carrier cyclic and so-called flex-deck operations to the hilt.[22] Marshaling these capabilities for an anticipated five-day surge option, CVW-5 proved itself able both to meet its CAS tasking and to provide strike packages for attacks against fixed targets in Baghdad. The S-3s of CVW-5 were crucial in making that flexible employment possible.[23]

The other two air wings in the Gulf followed a similar plan to enhance their capabilities. It turned out that S-3 organic tankers in the Gulf-based air wings were sufficient in number and capability to permit a release of some Air Force tankers allocated to *Abraham Lincoln* and *Constellation* to support other strikes emanating from the Gulf region, including from *Kitty Hawk*, as well as Air Force heavy bombers arriving from Diego Garcia and elsewhere. This concept had been tested and validated during the final days of Operation Southern Watch, during which organic tanking of outbound and returning carrier-based aircraft allowed F/A-18s to fly two consecutive 1+30 cycles on the same mission with little difficulty, with S-3s launching and recovering immediately before and after the refueling evolutions.

---

[22] Flex-deck operations, a more frenetically paced activity than normal cyclic operations involving waves of launches and recoveries at carefully scheduled intervals, typically have at least one of the carrier's two bow catapults firing continually, while the two waist catapults in the landing area are kept clear so that a steady stream of recovering aircraft can be accommodated.

[23] The S-3, which had its origins in the early 1970s as an antisubmarine warfare (ASW) platform, ultimately shifted over to the recovery tanking role after the last of the Navy's KA-6 organic tankers was retired from service and the open-ocean Soviet submarine threat to the carrier force largely went away with the collapse of the Soviet Union in 1991. The S-3's carrier-based ASW functions were subsequently taken over by the SH-60 multimission combat helicopter.

It also allowed an increase in on-station time for carrier-based fighters conducting CAS missions in southern Iraq around Basrah and Nasiriya. In so doing, it showed that a five-day surge capability could be sustained on an open-ended basis throughout the days to come.

The Super Hornet's Combat Debut

The Navy's latest combat aircraft to achieve operational status, the F/A-18E Super Hornet, entered fleet service in July 2002 when VFA-115 with CVW-14 deployed for the first time with the new aircraft in *Abraham Lincoln*. That aircraft was first developed and selected for series production during the early 1990s to redress a number of assessed capability and performance deficiencies of the basic F/A-18A and C that had been identified during the course of the latter's operational testing and evaluation, including range and endurance, maximum mission payload, usable munitions and external stores, bring-back capability (the total weight of munitions and stores with which the aircraft can be safely recovered aboard the carrier), survivability, and growth potential. The Super Hornet's assigned missions include maritime air superiority, air combat, reconnaissance, fighter escort, defense suppression, day and night strike, and through-the-weather target attack with satellite-aided JDAMs. Two upgraded General Electric F414-GE400 engines deliver 44,000 lb of combined thrust in full afterburner. That engine, with a nine-to-one thrust-to-weight ratio, offers 35 percent more power than that of the F/A-18C's F404 from which it was developed.[24]

The Super Hornet's larger wing and greater fuselage volume increase its internal fuel capacity by 33 percent (or 3,600 lb) over that of the C-model. In a high-altitude mission profile, the aircraft offers a 40 percent increase in range and an 80 percent increase in loiter time over that of the F/A-18C. The aircraft can remain unrefueled on a CAP station 200 nautical miles away from the carrier for 80 percent

[24] Richard R. Burgess, "Super Hornet Tallies 1000 Hours; AMRAAM Shot Successful," *Sea Power*, July 1997, p. 27.

longer than a C-model can. For strike missions, the E and the two-seat F variant, depending on configuration and weapons load, can fly 35–50 percent farther to a target than earlier F/A-18 models. That capability allows either strike missions deeper into enemy territory or flight operations conducted farther away from the nearest shore, or both. The Super Hornet also can carry a payload of 17,500 lb, about 25 percent more than the C-model's maximum load.

The new aircraft received a thorough combat shakedown in Operation Iraqi Freedom as VFA-115 played an important part in the war's initial combat operations. Later, more than two weeks into the war, VFA-41 with the first 14 F/A-18F two-seat Super Hornets arrived in theater in CVW-11 aboard *Nimitz*, which had been dispatched to the war zone to relieve *Abraham Lincoln.* During Iraqi Freedom strike operations, the Super Hornet averaged loads of up to 8,000 lb of air-to-ground ordnance and three air-to-air missiles for self-protection. The Super Hornet's increased payload capability over that of the earlier F/A-18C meant that the aircraft could attack the same number of targets (or more) while exposing fewer aircraft to enemy threats.

Although its maximum load at launch was later reduced to 4,000 lb because CAOC mission needs called for no more than that, the Super Hornet's increased bring-back capability was repeatedly demonstrated during on-call CAS and time-sensitive target attack operations, when orbiting aircraft never received target assignments from the CAOC. The Super Hornet could bring back to the carrier what the C-model would have had to jettison into the sea. In all, more than 350,000 lb of ordnance were dropped by VFA-115's F/A-18Es during the three weeks of major combat in Operation Iraqi Freedom. In the course of achieving that result, the squadron averaged more than 55 flight hours a day.

Super Hornets also flew more than 400 tanker sorties during Iraqi Freedom, helping to compensate for a shortage of available U.S. Air Force and RAF tanking at times. In its organic tanker role, the aircraft transferred 3.2 million pounds of fuel in all to carrier-based strike aircraft during its deployment and 2.3 million pounds during the three weeks of major combat in Iraqi Freedom. (Four VFA-115

F/A-18Es were configured for use primarily as tankers.)[25] Throughout Iraqi Freedom, one aircraft out of the squadron's dozen was fully dedicated to tanking, although CVW-14's S-3s retained the recovery tanking role. As the war progressed, Super Hornets provided organic tanking for carrier-based strike fighters all the way to Baghdad and back.[26] Since the aircraft, even in its tanker role, was equipped with a self-protection capability that included AIM-9 and AIM-120 air-to-air missiles and an electronic warfare (EW) suite for protection against surface-to-air threats, it could accompany strikers into defended airspace over enemy territory and also could help escort damaged aircraft back to friendly airspace if needed.

With respect to maintainability, the F/A-18E experienced the fewest maintenance man-hours per flight hour of the entire air wing in *Abraham Lincoln* throughout Iraqi Freedom's three-week period of major combat. At the start of the war, the squadron was averaging around 15 maintenance man-hours per flight hour, as compared to 20 for CVW-14's F/A-18Cs and 60 for its F-14s. Part of this greater ease of maintainability was a result of the squadron's ready access to electronic technical publications and its ability to receive technical manual updates electronically. (Part also emanated from the simple fact that the Super Hornet was a brand-new aircraft.)

The F/A-18F was the first Navy aircraft to be fully equipped with the Advanced Tactical Forward-Looking Infrared (ATFLIR) pod that is now in full-rate production.[27] Revealed problems with the initial ATFLIR sets that accompanied VFA-115 in *Abraham Lincoln*

---

[25] Hunter Keeter, "Tanking, 'Bring-Back' Highlights of Super Hornet's Performance," *Defense Daily*, June 2, 2003, p. 2.

[26] In all, a tanker-configured Super Hornet can carry more than 29,000 lb of fuel, including 480 gallons in each of four wing-mounted external fuel tanks and 330 gallons in an aerial refueling store mounted on the centerline station. During a short cycle, an F/A-18E/F in the tanker configuration can give away 25,000 lb of fuel. During a normal cycle, it can transfer more than 15,000 lb and more than 12,000 lb in a two-hour cycle. It is best used for recovery tanking, which nonorganic long-range tankers cannot provide, whereas the latter, with their substantially greater fuel reserves available for transfer, are preferable for en route refueling during extended-range combat missions.

[27] Carol J. Williams, "Super Hornet Creates a Buzz in the Gulf," *Los Angeles Times*, April 1, 2003.

were later overcome with the successor-generation of operationally ready pods that were provided to another F/A-18E squadron and to the first F/A-18F squadron embarked in *Nimitz*.[28] Reports from CVW-11 were that the ATFLIR's laser was deemed operational and that the pod's infrared and electro-optical images were of better resolution and quality than those produced by older ATFLIR development pods.

Highlights of the Carrier Contribution

The six carrier battle groups that participated directly in Operation Iraqi Freedom were the core of a larger U.S. naval presence in the war zone that included three amphibious ready groups and two amphibious task forces totaling nearly 180 U.S. and allied ships, 80,800 sailors, and another 15,500 Marines. At the high point of Iraqi Freedom, the Navy had deployed around the world eight carrier battle groups, eight big-deck amphibious ships, 21 combat logistics ships, and 76 sealift ships.

In all, more than 700 Navy and Marine Corps aircraft participated in Iraqi Freedom. Those combined assets contributed to a coalition total of 1,801 aircraft, 863 of which were provided by the U.S. Air Force.[29] Out of a total of 41,404 coalition sorties flown altogether, Navy and Marine Corps aircraft flying from carriers and large-deck amphibious ships flew nearly 14,000. Of those, 5,568 were fighter sorties, 2,058 were tanker sorties, 442 were E-2C sorties, and 357 were ISR sorties.

---

[28] The FY 2002 report of the Pentagon's Director of Operational Testing and Evaluation noted that ATFLIR was expected to correct the documented performance shortfalls in Nitehawk, including its problem with high-altitude magnification and resolution that degrades and, in some circumstances, altogether precludes even target location, let alone precise DMPI placement. The new system has demonstrated greater effectiveness than that of LANTIRN. (Christopher J. Castelli, "Enduring Freedom's Not-So-Secret Weapon: Navy F-14 LANTIRN Pods," *Inside the Navy*, April 29, 2002.)

[29] Lieutenant General T. Michael Moseley, USAF, *Operation Iraqi Freedom—By the Numbers*, Shaw AFB, S.C., Assessment and Analysis Division, Headquarters U.S. Central Command Air Forces, April 30, 2003, pp. 6-10.

Generating enough sorties to meet mission needs was never a problem. The six committed carriers and their embarked air wings could generate sorties faster than the CAOC could generate targets. The embarked air wings surged for 16-hour flying days for 23 days straight. Carrier air employment in Iraqi Freedom mainly featured two-cycle operations. FAC-A and reconnaissance sorties, however, were typically three- and four-cycle operations because of their longer duration. For the most part, the war featured larger strike packages and shorter-duration sorties for the Navy than did Operation Enduring Freedom.

Real-time targeting and precision strikes reached an unprecedented high in both numbers and intensity in Iraqi Freedom. More than 800 targets were attacked within the time-sensitive targeting process, with an average of 3.5 hours from target nomination to ordnance on target. More than 78 percent of the Navy's strike sorties received their target assignments in flight. Navy FAC-As provided 24-hour-a-day on-station service over southern Iraq and 16-hour-a-day service in the north. Strike-fighter operations also featured unprecedented flexibility in the selective use of satellite-aided JDAMs or LGBs, depending on assessed targeting needs.[30]

Of the 5,300 bombs dropped by Navy strike aircraft, fewer than 230 were unguided. More than 75 percent of the precision weapons delivered by Navy strike aircraft were JDAMs. As in Enduring Freedom, a considerable amount of strafing was also conducted by allied strike fighters. The air wing in *Truman* alone expended 20,000 rounds of 20 mm ammunition during the course of the war.[31]

Carrier air operations over northern Iraq were similar to those that largely predominated in Operation Enduring Freedom, in that they entailed a continuous "airborne presence with weapons," with

---

[30] Rear Admiral Matthew Moffit, USN, Commander, Naval Strike and Air Warfare Center, "Naval Aviation 2010–2020: A Decade of Transition," briefing to the Navy TACAIR Symposium 2004, n.d.

[31] Air wing commander panel on carrier air operations during Operation Iraqi Freedom at the Tailhook Association 2003 annual symposium, Reno, Nevada, *The Hook,* Fall 2003, p. 65.

friendly SOF units closely intermingled with the enemy and with ordnance bring-back a matter of course.[32] Operations in the south, in contrast, entailed more classic carrier air-wing strike missions in a target-rich environment and with a clearer separation of friendly and enemy forces on the ground. A typical mission flown from the Mediterranean over Turkey into northern Iraq would entail a night launch and an hour or more of transit time to the tanker rendezvous point, followed by entry into Iraqi airspace in two-plane sections or four-plane divisions and receipt of mission tasking from the CAOC while the aircraft were en route to their assigned targets or holding areas. These formations would return to the tanker track for inflight refueling two or three times during the course of a mission, followed by one last tanker contact after the window of aircraft vulnerability to enemy fire was closed or all ordnance was expended before returning to the carrier for a night approach and recovery. Such missions routinely lasted from five to seven hours. Night refuelings in bad weather and in heavy turbulence with low fuel states were the rule rather than the exception.

Because there was no fire-support coordination line (FSCL) associated with strike operations in support of SOF activities in northern Iraq, only kill box interdiction was conducted by allied fighters. The E-2C assumed the ABCCC role in these operations, coordinating with the CAOC to receive tasking, relaying tasking and contact information from tactical air control parties (TACPs) to shooters, passing along retasking and inflight mission reports, and managing tanking. The contribution of the carrier air wings was to maintain airborne weapons on call around the clock as far as 500-700 nautical miles away from the carrier operating areas in the eastern Mediterranean. This responsibility had the two air wings performing flight operations 15 hours a day with five to seven packages of eight to ten aircraft each in covering assigned windows of vulnerability to enemy fire. Early problems with these operations included a shortage of

[32] Robert W. Ward, Allen Hjelmfelt, Carter Malkasian, Daniel Roek, John Tand, and Daniel Whiteneck, "Operation Iraqi Freedom: CVW Fire Support to Ground Forces," undated briefing charts, Fairfax, Va.: Center for Naval Analyses.

available tanking, uneven interaction with TACPs because of poor communications, and spotty coordination with the CAOC and subordinate control entities in minimizing ordnance bring-back.[33]

Carrier-based E-2Cs performed well in the ABCCC role and were often indispensable in getting carrier-based strike aircraft to the right kill boxes. They also provided air traffic control and served as a communications link between the carriers, the CAOC, and ground tactical commanders. The EA-6B also played a pivotal part in the air war. As before during Operations Allied Force and Enduring Freedom, the availability of support jamming was an absolute go/no-go criterion for all strike missions, including those involving the Air Force's B-2 and F-117 stealth aircraft.[34]

The average flight operations day was 16 hours aboard each carrier during the first 23 days, after which it ramped down to around 13–14 hours. Each air wing averaged 120–130 sorties a day. Flight deck activity often continued without interruption for 24 hours a day for long stretches, since strike aircraft and tankers frequently recovered later than planned as a result of repeated CAS requests. Navy strike aircraft were airborne 24 hours a day, seven days a week. As in Operation Enduring Freedom, alert strike packages were launched every day as previously undiscovered targets of interest were identified.

Adequate LGB and JDAM stocks allowed the deployed air wings to play a significant role in seeking out and attacking any enemy ballistic missile launchers that might be discovered. The two wings operating out of the eastern Mediterranean focused on suppressing Scud launches into Israel from Iraq's western desert. The three Gulf-based wings concentrated on potential missile launches from southern Iraq against Kuwait, Saudi Arabia, and other neighboring Gulf states. Each carrier had around 40 magazines for munitions storage, with row after row of GBU-12 and GBU-16 LGBs

---

[33] Ward et al.

[34] Robert Wall, "E-War Ramps Up: EA-6B Prowler to Resume Traditional Radar-Jamming Role if Iraqi Conflict Escalates," *Aviation Week and Space Technology,* March 17, 2003, p. 49.

stacked up for use. It took around ten crew members roughly 10–12 minutes to build up a bomb ready for use.[35] Because the CAOC could not always guarantee targets that could be positively identified and attacked without an unacceptable risk of collateral damage, strike aircraft sometimes recovered with unexpended ordnance. A maximum of 2,000 lb of allowable bring-back ordnance was eventually dictated by the circumstances of the ground advance.

The war featured a number of firsts for carrier aviation as old systems more than a few times accomplished new things. For example, Operation Iraqi Freedom saw the first delivery of JDAMs by the F-14D, the first use of the EA-6B in a psychological operations role, and a laser Maverick missile fired in combat for the first time by an S-3.

On Balance

Throughout the 12 years that separated Operations Desert Storm and Iraqi Freedom, the Navy's carrier air wings underwent a steady improvement in their precision effects, lethality, and ability to participate seamlessly in joint operations. That improved capability was amply validated in Operation Iraqi Freedom. Even more so than the Afghan air war that preceded it by only a year, the Iraqi Freedom experience showcased a Navy that can surge its carrier air arm on demand with reasonable notice. Starting in August 2002, seven of the Navy's ten carrier air wings, with 488 embarked aircraft in all, were deployed and on call to participate in or indirectly support the looming war effort.[36] That performance showed the payoff of the more than $7 billion that the Navy had spent over the preceding three years on force readiness enhancements, including increased fly-

[35] Lyndsey Layton, "Building Bombs Aboard the Abraham Lincoln," *Washington Post*, March 14, 2003.

[36] Vice Admiral Kevin Green, USN, Deputy CNO for Plans, Policy, and Operations (N3/N5), "Operation Iraqi Freedom," briefing to Representative Roscoe Bartlett, Washington, D.C., n.d.

ing hours and additional provisions for ordnance, spare parts, and maintenance needs. Moreover, the training received by the most recently minted aviators in the newly standardized Strike Fighter Weapons and Tactics (SFWT) syllabus, a concentrated blend of academics, part-task training, emphasis on airborne execution, and thorough mission debriefing, minimized first-time combat mistakes in Operation Iraqi Freedom.

Even with the introduction of the Super Hornet into line service in just enough time for it to participate in the war, however, the F-14 remained a viable platform, since it offered great range and its LANTIRN targeting pod was available in greater numbers than the F/A-18E's new ATFLIR pod. Tactics, techniques, and procedures (TTPs) for the F-14 focused on hunting down mobile targets. As those TTPs were developed and refined, they registered major advances in finding objects of interest on the battlefield quickly. Navy mission planners soon realized that the JDAM offered little capability against moving targets on the ground. Provision of mensurated coordinates for attacking mobile targets was not possible because friendly ground forces lacked the technical wherewithal for generating such refined coordinates. The F-14's full-capability TARPS and FTI, however, enabled real-time imagery transfers whereby SOF teams on the ground could cue F-14s to locate and attack moving targets with LGBs.[37]

The Iraq war also set a new record for close Navy involvement in the high-level planning and command of joint air operations. According to then–Vice Admiral Timothy Keating, the Combined Force Maritime Component Commander (CFMCC), detailed and coordinated planning had taken place beforehand during the buildup for war between the Navy's 5th Fleet and Central Command Air Forces (CENTAF) in determining which personnel the Navy would send to the CAOC and what their qualifications needed to be. If a qualification requirement were not met, selected individuals received rush schooling in the needed skills to enable them to augment the

[37] Lieutenant Commander Richard K. Harrison, USN, "TacAir Trumps UAVs in Iraq," *Proceedings*, November 2003, pp. 58–59.

CAOC staff. Admiral Nichols, the deputy CFACC, had an augmentation team of 101 Navy personnel, around half of whom were reservists and the rest of whom were drawn from the weapons schools and fleet units. That made for a visible and influential Navy presence in key CAOC leadership positions.

Perhaps most important of all, Operation Iraqi Freedom was a true joint and combined effort in which all force elements played equally influential roles. Admiral Keating characterized the operational payoff of all this as "joint warfighting at the highest form of the art I'd ever seen. . . . There was understanding, friendship, familiarity, and trust among all the services and special forces working for General Franks. He did, in my view, a remarkable job of engendering that friendship, camaraderie, and trust. In fact, he insisted on it. . . . There was no service equity infighting—zero."[38]

---

[38] Interview with Vice Admiral Timothy J. Keating, USN, "This Was a Different War," *Proceedings*, June 2003, p. 30.

CHAPTER FOUR

A New Carrier Operating Concept

Before the terrorist attacks of September 11, 2001, the Navy's worldwide "presence" posture had been enabled and supported by a highly routinized and predictable sequence of maintenance, training, and unit and ship certification aimed at meeting scheduled deployment dates that were all but set in stone. The sudden demands levied on the Navy by the events of September 11, however, changed that familiar pattern of operations irretrievably. Recognizing that the emergent demands of an open-ended global war on terror meant a need for a more responsive naval force able to sustain a higher level of mission readiness, the CNO, Admiral Vern Clark, in March 2003 announced a need for the Navy to develop a new Fleet Response Concept (FRC) aimed at creating and institutionalizing a new approach to readiness that would provide the Navy with an enhanced carrier surge capability. The CNO further tasked the commander of Fleet Forces Command, Admiral Robert Natter, to develop and implement this new operating concept, which drew on the preceding two years of heightened fleet readiness, including the still-fresh experience of Operation Enduring Freedom.

By a convenient stroke of good timing, this FRC initiative was put into effect on the eve of Operation Iraqi Freedom, which provided an ideal test of the concept under fire. As the Iraqi Freedom air war neared, the Navy had eight carrier battle groups and air wings deployed worldwide, including USS *Carl Vinson* and her embarked CVW-9 in the Western Pacific covering North Korea and China during the final countdown. Five of those eight battle groups and air wings had participated in Operation Enduring Freedom just a year before. With five carrier battle groups on station and committed to the impending war, a sixth en route to CENTCOM's AOR as a timely replacement for one of those five, a seventh also forward-deployed and holding in ready reserve, and yet an eighth carrier at sea and ready to go, 80 percent of the Navy's carrier-based striking power was poised and available for immediate tasking.[1] During the cold war years, having eight out of 12 carriers and ten air wings deployed at sea and combat-ready at the same time would have been all but out of the question.

The extent of the expected demand for carrier-based air power in the impending war against Iraq presented an almost tailor-made opportunity for the Navy to exercise in a real-world setting its new approach to fleet operations. Indeed, Operation Iraqi Freedom, for which the Navy deployed no fewer than 70 percent of its ships, epitomized the new requirements for which the FRC had been developed. Immediately after the fall of Saddam Hussein's regime, seven carrier battle groups were conducting forward operations, with an eighth (USS *Nimitz*) about to join them.[2] This experience taught the Navy that the methodology used throughout the cold war for manning, maintaining, and training the fleet would not produce, on a sustained and open-ended basis, the carrier surge readiness that the new security environment of the 21st century required.[3] Accordingly, shortly after the Navy's successful contribution to Iraqi Freedom, Admiral Clark approved Admiral Natter's now-validated FRC proposal and directed its implementation as the Fleet Response Plan (FRP), a fundamentally new operating approach that essentially aimed at making the most of smarter and more efficient resource-

---

[1] Vice Admiral Michael D. Malone, USN, "From Readiness at Any Cost to Cost-Wise Readiness," *The Hook*, Summer 2004, p. 5.

[2] Admiral Vern Clark, USN, "Persistent Combat Power," *Proceedings*, May 2003, pp. 46–48.

[3] Coordinated fleetwide message on FRP from the commander, U.S. Fleet Forces Command and commander, U.S. Pacific Fleet, April 24, 2004.

management practices to provide carrier forces that can deploy on shorter notice and in greater numbers than before.

How the Surge Concept Works

The FRP formula seeks to increase the efficiency of maintenance and training processes and procedures so as to heighten overall carrier availability and readiness and to increase the carrier force's speed of employment.[4] At bottom, it envisages the augmentation of deployed carriers and embarked air wings with surgeable carriers and air wings that can be rapidly readied for deployment and combatant-commander tasking, thereby yielding increased overall force employability and an earlier commitment of carrier striking power. The CNO's Sea Power 21 vision is based on a concept of operations that aims, through FRP, to increase the Navy's force-projection capability from the former 12 full-up and standardized carrier *battle* groups and 12 amphibious ready groups to 12 more mission-specific carrier *strike* groups (CSGs) and 12 expeditionary strike groups (ESGs). In implementing FRP, the Navy will no longer focus narrowly on scheduled deployment dates but rather will concentrate on being able to "scramble" at least five or six carriers and their assigned air wings on short notice in response to contingencies that might suddenly arise.

More specifically, the FRP aims to provide combatant commanders with what has come to be characterized as "six-plus-two" ready CSGs. The "six" refers to deployable CSGs that can respond almost immediately to tasking, wherever they may be in their respective training and workup schedules, in varying amounts of time up to 30 days. The remaining two represent near-combat-ready CSGs that can deploy as needed on a more accelerated schedule than before, normally within around 90 days. That will make for a larger overall naval air force complement able to respond to tasking, as opposed to

[4] "Fleet Response Plan: Postured to Deter, Poised To Strike," briefing prepared by OPNAV N3/5 (Deep Blue), Office of the Chief of Naval Operations, Washington, D.C., March 2004.

a smaller forward-deployed force fielded at the highest readiness level solely to meet "presence" requirements.

In this new surge plan, the Navy will still aim as a matter of course to adhere to its previously established carrier deployment pattern of six months at sea followed by 18 months at home for maintenance and reconstitution, and it will depart from that pattern only in case of a major contingency need. However, Admiral Clark has indicated that even in peacetime, regular six-month deployments need not invariably be the norm. On the contrary, as the CNO put it, "you may be in a surgeable window, but that doesn't mean that you're going to surge." FRP, he explained, is not just about surging, but also about rethinking the lengths of deployments should a normal six-month cruise not be the most appropriate plan: "We should be less interested in [simply] presence," he said, "and more interested in presence with a purpose. If it is about presence with a purpose, it might just be a four-month deployment. I would rather muster two [strike] groups for three months and do something really significant internationally than just go over and hang out for six months without purpose, just kind of marking time."[5]

In this new scheme of operations, a CSG deployment will occur somewhere within a surge window, with the actual deployment date being determined by the JCS to meet declared combatant-commander needs. For the near term, ready CSGs will continue to fulfill combatant-commander "presence" requirements, but those deployment requirements will no longer be the only consideration that determines the readiness of a CSG. What remains unchanged in FRP is the planned *routine* deployment length of six months, a steady flow of personnel to and from sea duty, and the CNO's operational and personnel tempo goals.

FRP also aims to shorten the normal 18-month reconstitution process following a CSG's return from a deployment by accelerating maintenance schedules, changing personnel management procedures, and altering training rotations so that the group could be ready to

---

[5] Clark, "Persistent Combat Power."

deploy again after only nine months. Once the ships and the air wing are declared ready for deployment again, the CSG will then go through phases of "surgeability," being able to deploy as necessary at any moment in the six-month window before a regularly scheduled deployment.

In cases of the most immediate and demanding operational need, regional combatant commanders around the world can expect to gain emergency naval surge units that will come at a heightened level of risk because of the shorter time available for focused training to meet surge demands. That increased risk level could include, in the worst case, a reduced manning of some carriers and embarked squadrons, as well as a predeployment air-wing training effort that concentrated on achieving highly specific and necessarily limited readiness goals at the expense of certain other important proficiency categories in the interest of attaining a surge capability as soon as possible. That calculated risk should be acceptable to concerned commanders, however, since the later-deploying CSGs will presumably already enjoy the regional access established by previously arriving forces and can accordingly focus on achieving an acceptable readiness level in such critical mission areas as strike and CAS at the expense of, say, maneuvering air combat proficiency, which would require additional training time but might well not be needed.

At the heart of this new approach is what the Navy calls "targeted readiness." Instead of assuming a need to achieve the highest state of readiness as a precondition of any CSG's deployment, as was the norm for carrier battle groups throughout the cold war, it recognizes, in the words of a fleetwide message on FRP to all commanders, that "in many instances, absent indications of imminent danger or war, intermediate levels of readiness are not only acceptable but a prudent use of resources."[6] The core issue here concerns how much risk commanders will be willing to tolerate in accepting that slightly lower readiness level and, worse yet, possible readiness levels even lower than C2 (with C1 representing peak readiness, which was the

[6] Coordinated fleetwide message on FRP from the commander, U.S. Fleet Forces Command and commander, U.S. Pacific Fleet, April 24, 2004.

rock-bottom precondition for any carrier battle group deployment throughout the cold war). For example, almost surely the seventh and eighth CSGs and their attached air wings in a surge window, for example, will not be completely at C2 readiness.

In the case of more demanding contingencies, strike-fighter, tanker, and electronic attack aircraft and aircrews could also be drawn, as needed, from off-cycle air-wing squadrons. Thanks to recent fleetwide moves to standardize tactics, techniques, and procedures through the SFWT program, almost any F/A-18s from similar production years and any line pilots (including those drawn not just from other fleet squadrons but also from fleet replacement squadrons and NSAWC) could augment deployed squadrons with a minimum of needed predeployment workup training.[7] With respect to the deployed force, the desired manning goal is a number of skilled personnel adequate to enable a surge in air operations for four to five days, to include 24-hour-a-day operations for a single CSG. (The most critical skill sets in this regard reside in the aircrew complement, squadron maintenance, and the carrier's operations and air departments.)

In FRP, a baseline 44 strike-fighter air wing is projected as being able to conduct a nominal 150 sorties a day without augmentation. If supplemented by additional pilots, maintainers, and carrier air and operations department staff, it should be able to generate up to 250 sorties a day—and with double or triple the previous number of DMPIs that can be successfully attacked per sortie, thanks to the advent of precision-guided munitions as the standard weapons load on all strike fighters. Moreover, owing to the high premium the Navy placed during the 1980s on incorporating high-reliability features into its new aircraft designs, today's air wings experience substantially lower aircraft failure rates during normal operations than ever before. As a result, aircraft systems are no longer the limiting factor in the generation of a carrier air wing's striking power that they once were. In addition, these capability improvements mean that the proper

[7] Commander Donald H. Braswell, USN, "Carrier Strike Groups Need a Combat Surge Capability," *Proceedings*, September 2004, p. 77.

metric for measuring a carrier air wing's offensive power is no longer the *number* of sorties it can generate during a given period but rather the number of sorties it can generate by specific mission type and the *effectiveness* of those sorties in achieving a combatant commander's declared objectives.[8]

To meet the demanding goals of FRP, new approaches are required toward manning, training, and maintenance, with maintenance provisions configured to achieve a surge capability as soon as possible in the workup cycle. Training will now focus on acquiring designated core competencies earlier in the workup cycle in the process of meeting surge requirements. Contingency needs could dictate longer-than-normal deployment lengths for some surged CSGs depending on the character and demands of the operational challenge. The peacetime goal, however, will continue to be a firm deployment policy of six months, a minimum two-to-one turnaround ratio, and at least half of a ship's time spent in home port.

To support that goal, personnel assignment practices will need to be readjusted to ensure that deploying units have the personnel they will need to meet surge demands. As Admiral Natter put it, an important result of the FRP will be "a well-maintained, well-manned, and well-trained force that routinely deploys at *peak* readiness, is able to surge at a *high level* of readiness, and is capable of an emergency surge with an *adequate* level of readiness that minimizes risk."[9] Its goal, presence with a purpose, aims for *predictable* rather than unprogrammed risk.

Although the full extent of FRP's impact and potential has yet to be determined, it has become increasingly clear that the Navy, like all of the other services, has been asked by the nation's civilian leadership to do ever more with its existing assets. On this point, while he was still commander of Naval Air Force, U.S. Atlantic Fleet, then-

---

[8] Moreover, as the air portion of both Operations Enduring Freedom and Iraqi Freedom attested, the number of sorties generated will depend, at least at first, not on an air wing's sortie-generation capability but rather on operational circumstances and the CFACC's target-coverage needs.

[9] Admiral Robert J. Natter, USN, "Creating a Surge Ready Force," *Proceedings*, September 2003, p. 57, emphasis added.

Rear Admiral Zortman described the underlying idea of FRP as having "more of the fleet being ready to go more of the time," so that it might be configured to operate on a continuing basis in much the same way that it performed so effectively during Operations Enduring Freedom and Iraqi Freedom. However, he added, "there is a cost to FRP. We have to identify those costs [in] equipment, training, [and] people."[10]

The Payoff for Combatant Commanders

In essence, the objective of FRP is to double the number of carriers available at any time for mission tasking worldwide. Instead of having one carrier each in the Pacific Ocean, the Persian Gulf, and the Mediterranean Sea, FRP will enable the Navy to have up to *six* carriers ready to respond anywhere as a group within 30 days, with another two capable of deploying within 90 days if need be. This arrangement will enable the nation's carrier air assets to be more flexible and unpredictable. Viewed another way, FRP aims to have more of the carrier force ready for a longer stretch of time than before, with somewhat reduced readiness at the tail end of the carrier inventory at any given time. Unless contingency needs demand otherwise, however, the carrier strike-group utilization rate should remain roughly the same as before.

As demonstrated by its response with six carriers and their embarked air wings for Iraqi Freedom, the Navy now fights as a battle *force* rather than as individual carrier battle *groups*. Its challenge is to mount a more responsive and employment-capable force available to surge and sustain yet also able to be reset quickly through periodic refresher training and maintenance of ships, aircraft, and equipment. The need to generate *goal-oriented* presence around the world on short notice as contingency-response needs arise dictates, among other things, dispersed naval forces equipped with a mix of capabili-

[10] Sandra I. Erwin, "Naval Aviators Experience Success in Iraq, But Worry About the Future," *The Hook*, Fall 2003, p. 70.

ties, including wide-area sensing, command and control, and precision strike, to carry out a broad spectrum of operations.

On the premise that one learns by doing, FRP will very likely remain, in effect, a real-world battle-lab experiment for a time as the Navy continues to seek the skill sets and resource-management practices that can best be counted on to ensure that the carrier deployment pattern that was demonstrated to such impressive effect in Operations Enduring Freedom and Iraqi Freedom can be sustained indefinitely. Along the way, the Navy's leadership will need to remain mindful of the readiness and capability investments required to make FRP work as the latter continues to be further refined and established as a new way of life. Ultimately, the goal of FRP is to provide the nation with a more flexible and less predictable CSG deployment capability. As Vice Admiral Zortman, now commander of Naval Air Force, U.S. Pacific Fleet, and the Navy's lead type commander for aviation, has summarized this important goal, its essence is to offer the nation's leadership a carrier force that is able to "surge when necessary and throttle back when appropriate."[11]

Two downside considerations, however, could ultimately work to undermine the long-term feasibility and promise of FRP, perhaps even irretrievably. To begin with, because the Navy does not define or control when such sought-after throttling back is "appropriate," the resource tension between near-term readiness and longer-term force recapitalization is likely to become increasingly felt by the Navy in the coming years in the continued absence of funding relief for these competing needs. Second, and potentially even more consequential for the viability of the Navy's carrier surge goal, FRP was conceived and formulated on the core premise that the service would retain its long-standing post-cold-war force structure of 12 deployable carriers and ten active air wings. Since that concept was implemented in the early aftermath of Iraqi Freedom, however, the Department of Defense has, in a cost-cutting budget drill, moved with determination to reduce the Navy's carrier force from 12 to 11 carriers by retiring

---

[11] Comments during a flag panel discussion at the Tailhook Association's 2004 annual symposium, Reno, Nevada, September 11, 2004.

the conventionally powered USS *John F. Kennedy* within the coming year, 13 years before that carrier's slated decommissioning, to help pay for global contingency operations and draw down the federal deficit. The Office of the Secretary of Defense has further indicated of late that it is seriously considering an even further reduction of the Navy's carrier force below the currently envisaged 11 carriers.[12] Although detailed analysis of the implications of such radical cost-cutting measures for the surge potential of FRP has yet to be completed by the Navy, the service's leadership currently insists that its desired "six-plus-two" surge capability can be sustained even with only 11 deployable carriers and ten air wings.[13] However, that leadership also appears to be coming increasingly to the reluctant but inevitable conclusion that any further cuts in carrier and air-wing end strength below 11 carriers and ten air wings would not only make the "six-plus-two" surge goal unattainable as a practical option but might well render FRP unworkable altogether.[14]

---

[12] See Jason Sherman, "Defense Department Eyes Further Cuts to Aircraft Carrier Fleet," *Inside the Pentagon*, April 21, 2005, p. 1.

[13] The since-retired CNO himself, Admiral Clark, insisted that the Navy should be able to absorb the impending loss of USS *John F. Kennedy* with little adverse influence on its overall striking power because of the changes in the way the service deploys and maintains its carriers under the FRP. (Dale Eisman, "Senator Warner Opens Battle to Maintain Norfolk Fleet," *Norfolk Virginian-Pilot*, February 11, 2005.)

[14] Conversations with staff officers in OPNAV N78 and N81, Office of the Chief of Naval Operations, the Pentagon, Washington, D.C., April 19–20, 2005.

CHAPTER FIVE

The Next-Generation Carrier

The Navy's large-deck, nuclear-powered *Nimitz*-class aircraft carriers have provided the nation with effective force-projection service for more than 30 years. The baseline design for those carriers was completed during the 1960s, with a view toward replacing the Navy's five aging oil-fired *Forrestal*-class carriers on a one-for-one basis. Since then, the U.S. carrier force has not even once undergone an aggressive research and development (R&D) effort to fold cutting-edge technology into the design of a follow-on ship. Instead, from USS *Dwight D. Eisenhower* (CVN-69) onward, all of the Navy's aircraft carriers have been "modified repeats," if not carbon-copy repeats, of previous carriers. Carrier flight-deck design has also remained static for decades. The last completely new deck design that was incorporated into the *Nimitz* class of carriers was little more than an add-on to a variant of *Forrestal*-class carrier decks dating back to 1955.

In light of these considerations, and prompted by growing concern that the continued absence of any significant change in U.S. carrier design was inhibiting operational capability improvements and the incorporation of new technologies, the Navy in 1993 commissioned a future sea-based air platforms working group to explore operational requirements, available technologies, and needed R&D initiatives to ensure that the next generation of carriers would reflect essential elements of the ongoing revolutions in military and business affairs. That initiative led, in turn, to the Navy's issuance in 1996 of its first mission-needs statement for a 21st-century sea-based strike aircraft platform to replace *Enterprise* and the early *Nimitz*-class carriers beginning in 2013.[1] That statement appeared in a document that resulted in the establishment of the Future Aircraft Carriers (CVX) program.

At the onset of this initiative, it was anything but a foregone conclusion that the new carrier class would be provided with nuclear rather than less costly conventional propulsion. Debate also persisted for a time over the option of developing a new class of smaller *Midway*-size carriers in lieu of CVX. To help weigh the pros and cons of these competing alternatives and adjudicate them from a comparative operational effectiveness perspective, the Assistant Secretary of the Navy for Research, Development, and Acquisition in January 1997 tasked the Naval Research Advisory Committee (NRAC) to conduct a study of the science and technology opportunities that might best affect the engineering and operational flexibility of CVX. The NRAC study, co-sponsored by the Deputy Commander for Engineering of the Naval Sea Systems Command and the Director of Air Warfare in the Office of the CNO, eventually concluded that CVX should be designed to support a large (80-aircraft) air wing and to be able to conduct flight operations in heavy sea states to undertake any assigned mission.

One of many challenges identified by the study panel in this regard was to ensure that CVX could be operated normally with a smaller air wing, yet surged as needed to a larger air wing as the most stressing short-notice missions might dictate. The study further concluded that CVX must be able to undertake any assigned mission under all but the most prohibitive weather conditions, requiring the best possible sea-keeping capabilities, and also must be available for task-

---

[1] Because the Navy's first nuclear carrier USS *Enterprise* has eight reactors whereas the newer *Nimitz*-class carriers have only two, the engineering departments of the latter require 250 fewer personnel. Before the decision was made in 1992 to cut the U.S. carrier force structure from 15 to 12 carriers, the Navy had already opted to refuel rather than retire *Enterprise*, thus extending her service life by 20 years. In 1994, Newport News Shipbuilding completed refueling Enterprise's eight smaller reactors at a cost of $1.5 billion over four and a half years. (Richard R. Burgess, "Newport News Gears Up for CVN Refueling," *Sea Power*, September 1997, p. 31.)

ing anywhere and anytime, dictating a high-speed sprint capability. Those desiderata, in turn, were found by the study to require a nuclear propulsion plant and a displacement on the order of 100,000 tons. On balance, as reported in its Executive Summary, the study panel "found arguments favoring large, nuclear-powered carriers to be persuasive when considered in the light of reduced overseas bases and the wide spectrum of rapidly-evolving crises which are likely in the 21st-century environment."[2]

A subsequent study commissioned by the Navy's CVX program office in 1998 sought to explore the differences in military effectiveness of a future CVX design propelled by nuclear and nonnuclear power. The comparative life-cycle costs of the competing alternatives were not a consideration in this assessment, although the study acknowledged at the outset an independent Navy estimate that the nuclear-propulsion alternative would be roughly 10 percent more expensive over time than a gas-turbine alternative, even though *both* designs would have a lower life-cycle cost than that of the *Nimitz*-class carriers that either CVX alternative would replace. The study further noted that the cost-estimating risk attached to the conventional alternative was higher than for the nuclear alternative, since the cost and availability of fossil fuel over the next half-century cannot be estimated with great accuracy. It also noted that the gas-turbine option is basically unproven for a ship the size of a large-deck aircraft carrier, whereas the nuclear-powered alternative would entail an extension of proven propulsion technology that would not be affected by oil pricing and other uncertainties.[3]

After examining the relative operational effectiveness of the two competing propulsion alternatives in both single-carrier scenarios and high-tempo surge operations involving four CVX strike groups oper-

---

[2] William F. Weldon and others, *CVX Flexibility*, Office of the Assistant Secretary of the Navy for Research, Development and Acquisition, Naval Research Advisory Committee Report NRAC 97-1, Washington, D.C., October 1997, p. 7.

[3] Many civil-registered cruise ships have been successfully powered by gas-turbine engines, but the heaviest of those have grossed out at far less than the roughly 90,000 tons that the typical *Nimitz*-class carrier weighs.

ating in the three principal global carrier operating hub areas (the Mediterranean, Southwest Asia, and the Western Pacific), the study found that the nuclear-powered CVX would be more effective than a gas-turbine counterpart in all conceivable circumstances. It also found no evidence of any operational (as opposed to cost-savings) benefit from adopting gas-turbine propulsion for CVX. On the contrary, it found in every case that the nuclear-propulsion alternative had proven itself superior in flexibility, combat capability, and rapidity of response to crises occurring in the operational hubs. Not only did the study find no comparative operational advantage to be had from relying on gas-turbine propulsion, it determined that the latter's reliance on refueling every few days to remain operationally ready was "a significant drawback, not only for the aircraft carrier itself, but for the battle group and the Navy's refueling force as a whole." Indeed, it concluded, should the conventional propulsion alternative be selected for CVX, "the now-hackneyed phrase 'Where are the carriers?' could well become 'Where are the oilers?'"[4]

In the end, largely on the strength of such assessments, the Defense Acquisition Board (DAB) in June 2000 approved the Navy's proposed plan to pursue a follow-on to the *Nimitz*-class carrier that would be a large-deck and nuclear-powered ship to be called the CVNX. That approval came on the heels of a four-year internal Navy analysis of alternatives, plus three high-level reviews by the Office of the Secretary of Defense.[5] The DAB's consensus was that large-deck carriers were the preferred alternative. The Navy's analysis of the range of choices also concluded that opting for nuclear rather than conventional power would add approximately 10 percent to a new carrier's life-cycle cost irrespective of the carrier's size, but, ultimately nuclear rather than conventional power was chosen for the next-generation carrier.

[4] Roger N. Whiteway and Thomas Vance, *The Military Effectiveness of CVX Propulsion Alternatives*, McLean, Va.: Science Applications International Corporation, September 1998, p. 5-3.

[5] David A. Perin, "Are Big Decks Still the Answer?" *Proceedings*, June 2001, pp. 30–33.

Although those two choices in close succession narrowed the spectrum of potential change in the CVNX design process, they did not settle all cost and performance issues that still awaited decision. At that time, still-pending cost-savings considerations included such possible measures as decreasing the size of the flight deck and reducing the carrier's top speed. On the survivability front, they included reducing the number of defensive systems and certain armor protection options in the "austere-ship" category. Improved survivability measures were also considered for the "robust-ship" alternative, such as advanced armor protection, signature reduction, and state-of-the-art damage control systems.

For a time, cost considerations maintained the upper hand in this debate, rendering the "austere" approach to CVNX a real possibility until the Enduring Freedom experience in late 2001 and early 2002 appeared to empower a marshalling of compelling arguments against it. Even then, however, the next-generation carrier program was not yet out of the woods. In May 2002, Secretary of Defense Rumsfeld issued a Defense Planning Guidance (DPG) document for the upcoming Future Years Defense Program (FYDP) for FY 2004 to FY 2009 that directed a reexamination of the nation's need for five major defense acquisition programs, including what by that time had been defined by the Navy as CVNX-1, the first of two planned evolutionary ship designs that would bridge the *Nimitz* class and the next generation of carriers.[6] In response to this directive, the Office of the Secretary of Defense (OSD) began considering several alternatives to the Navy's existing carrier force modernization plan, including procuring smaller conventionally powered carriers instead of nuclear-powered carriers, opting for a repeat version of CVN-77 (the last of the planned *Nimitz*-class carriers) in FY 2007 rather than proceeding

---

[6] The other four programs were the Air Force's and Marine Corps' V-22 Osprey tilt-rotor troop transport aircraft; the Air Force's F/A-22 next-generation fighter; and the Army's RAH-66 Comanche attack and reconnaissance helicopter and Crusader heavy artillery piece. Not long thereafter, Crusader and Comanche were cancelled outright. See Pat Towell, "Crusader May Be Precursor to More Defense Cuts," *Congressional Quarterly Weekly*, July 20, 2002, p. 1963.

with CVNX-1, and skipping the procurement of CVNX-1 altogether.[7]

By way of background to this then-envisaged first successor to the *Nimitz* class of carriers, the Navy's initial inclination was to design its next-generation carrier directly "from a clean sheet of paper," as many described the service's desired approach at the time. In June 1998, however, the CNO, Admiral Jay Johnson, directed a more cautious and evolutionary—or "spiral"—approach over the course of construction of three successive new carriers, with the first of these (CVN-77) viewed as a transitional platform intended to serve as a way station to CVN-78, the first of the next-generation carriers to be launched in 2013, followed in turn by the more revolutionary CVN-79, which would fulfill the new carrier design.[8] In that envisaged plan, CVN-78, designated CVNX-1 at the time, would be a further-evolved version of CVN-77 that incorporated only a portion of the innovations planned for the next-generation carrier, including a reduced crew complement with 350 fewer members, a new and less costly nuclear power plant, a new electrical power distribution system, and electromagnetic rather than steam-powered aircraft catapults. In turn, CVN-79, then called CVNX-2 and slated to enter service in 2018, would complete the transition to a successor generation of carriers, incorporating a further reduction in the ship's crew size, a substantially redesigned flight deck, hull design improvements, and an electromagnetic aircraft recovery system, among other innovations. As the Navy's program executive officer for carriers at the time, Rear Admiral Roland Knapp, subsequently explained in 2001, this more evolutionary approach would enable the service to transition to the next generation of carriers at only a fraction of the buy-in cost that would have been required to leap directly from the reactor and

[7] Ronald O'Rourke, “"Navy CVNX Aircraft Carrier Program: Background and Issues for Congress," *CRS Report for Congress*, Order Code RS20643, Washington, D.C., Congressional Research Service, the Library of Congress, November 14, 2002, p. 4.

[8] Richard Lazisky, "CVX: Evolution of a Carrier," *Sea Power*, October 1998, p. 51. The "clean sheet of paper" approach would have required an R&D investment of $7 billion before 2006, a funding requirement that fell outside the Navy's budget guidelines for FY 2000.

associated technologies developed for USS *Nimitz* in the mid-1960s to a new class of nuclear-powered carriers for the 21st century.[9]

In a partial endorsement of this planned approach, a study by the Defense Science Board commissioned on September 6, 2001, by the Under Secretary of Defense for Acquisition, Technology, and Logistics, E. C. Aldridge, Jr., and released to the public in an unclassified version in late October 2002, recommended that the Navy be allowed to proceed with CVNX-1 as planned but that the designs for CVNX-2 and beyond should be held in abeyance until available technologies were further assessed.[10] The Navy's planned evolutionary approach was subsequently scuttled altogether toward the end of 2002 at the direction of Secretary Rumsfeld, on the reported ground that OSD did not consider the baseline CVNX-1 design to be sufficiently "transformational."[11] In OSD's view, it made more sense to proceed directly to the full-up next-generation carrier design by incorporating into CVN-78 all of the additional design features that had been initially intended for CVNX-2. After the CNO presented OSD with a credible plan to enable the Navy's carrier program, in effect, to skip a generation, Secretary Rumsfeld signed a Program Decision Memorandum on December 12, 2002, that pulled forward into the first ship technologies that had initially been intended to be deferred to CVNX-2, at which point the CVNX program was renamed CVN-21. To accommodate an orderly incorporation of the additional technologies and design features, a revised acquisition strategy slipped the new carrier's planned start from FY 2006 to FY 2007.[12] The only major CVNX-2 design item not included in that

---

[9] Scott C. Truver, "Today's Carrier Force—and Tomorrow's," *Sea Power*, June 2001, p. 40.

[10] For the full report, see *Defense Science Board Task Force on Future of the Aircraft Carrier*, Office of the Under Secretary of Defense for Acquisition, Technology, and Logistics, Washington, D.C., October 2002.

[11] O'Rourke, "Navy CVNX Aircraft Carrier Program," p. 4.

[12] Robert M. Schatzel, CVN-21 design manager for integration, *CVN-21 Program Overview*, briefing presented to the author, Newport News, Va.: Northrop Grumman Corporation, May 18, 2005.

changed approach was a new hull design, which will now be deferred until the need for an improved design is explored more fully from a performance and cost-benefit perspective.

In this amended approach, the last of the *Nimitz*-class carriers, USS *George H. W. Bush* (CVN-77), will be commissioned in 2008. Construction of the first CVN-21 will begin concurrently, with the ship scheduled to join the fleet in 2015, just in time to replace the Navy's first nuclear carrier, USS *Enterprise*, which will be more than 50 years old by that time. The second CVN-21 remains slated for commissioning in 2019. Before the first CVN-21's commissioning, *George H. W. Bush*, the last of the *Nimitz* class and a modified repeat of *Ronald Reagan*, will incorporate integrated systems along with other marginal improvements that will be carried over into CVN-21.

Thanks to increased efficiencies through a reduction in workload and selected automation, the manning requirement for CVN-21 will be only 60 percent that of the *Nimitz*-class carriers. (The *Nimitz* carriers were developed and produced in an era of relatively inexpensive manpower and before the recent revolution in information technology.) CVN-21 will feature other radical innovations as well, including an advanced reactor design derived from technology already developed and proof-tested in the reactor that now powers the latest *Virginia* class of nuclear fast-attack submarines. Electrification of all auxiliary systems will increase the ship's electrical power-generation capability to three times that of the *Nimitz* design and also will replace steam and hydraulic piping throughout the ship. In addition, it will allow four electromagnetic aircraft launch system (EMALS) catapults to replace the earlier-generation steam catapults. CVN-21 will have a more efficient flight deck and advanced arresting gear for aircraft recoveries. It also will have a smaller island positioned farther aft and farther outboard on the flight deck. Its new flight deck will be able to generate as many as 150 more sorties a day than current carriers are capable of producing.

Although the number of strike aircraft in tomorrow's carrier air wings will be lower than that of today's, CVN-21's greater flight deck area will make it easier for aircraft to be moved about the deck and respotted.[13] In all, a hallmark of CVN-21's design goals is the provision of an adaptable infrastructure that will enable the incorporation of new capabilities as they develop, which should greatly reduce the carrier's life-cycle costs over its planned service life.[14] Just as some *Nimitz*-class carriers will remain in service until 2050 and beyond, some of the CVN-21-class will still be in service at the turn of the 22nd century.

---

[13] Norman Friedman, "World Naval Developments," *Proceedings*, July 2003, pp. 5–6.

[14] For additional details on CVN-21's planned design features and improvements on the various capabilities of the Navy's current *Nimitz*-class carriers, see Vice Admiral John Mazach, USN (Ret.) and Adam Siegel, "CVN-21 and Sea Power 21: A Program Aligned with the Navy's Transformational Vision," *Wings of Gold*, December 2003, pp. 14–15; and Adam Siegel and Robert M. Schatzel, "CVN-21 and HSI [Human Systems Integration]," *Wings of Gold*, April 2004, pp. 30–32.

CHAPTER SIX

The Changing Face of American Carrier Air Power

Among the many substantial gains registered in the combat leverage of American carrier air power over the past decade have been its proven ability to surge a large number of carrier strike groups (as many as eight out of 12 carriers and ten air wings) and to keep them on station for the duration of a major campaign; to generate and sustain as many sorties as might be needed to meet a CFACC's daily target-coverage requirements; to service multiple aim-points with consistently high accuracy on each combat sortie around the clock irrespective of weather; and—with the help of nonorganic Air Force and allied long-range tanker support—conduct deep-attack missions well beyond littoral reaches and maintain armed strike fighters on station for hours, if need be, in providing on-call interdiction and CAS. These are all fundamentally new carrier force-projection capabilities that would have been all but unthinkable during the waning years of the cold war.

In the decade to come, this process of evolutionary change in naval strike aviation will continue to develop in a way that portends a revolutionary improvement in the overall combat potential of U.S. carrier air power. In particular, the immediate years ahead will see a further sharpening of the combat edge of the F/A-18E/F Super Hornet, a successor-generation of naval electronic warfare aircraft in the EA-18G, the introduction of the E-2D Advanced Hawkeye offering significantly increased airborne surveillance and battle-management capabilities, the Navy's long-overdue entry into the era of all-aspect stealth with the pending introduction of the F-35C JSF, a leaner yet more efficient and capable carrier air-wing force structure, and a further shift from platform-centric to network-centric operations.

Planned Super Hornet Improvements

A succession of major aircraft program cancellations almost on the heels of one another during the first half of the 1990s put naval aviation into the force-modernization equivalent of a post-stall gyration that, for a time, seemed to threaten to devolve into a nonrecoverable flat spin. That programmatic departure from controlled flight, as it were, was finally broken when the F/A-18E and F Super Hornet joined the fleet in 2002. That long-awaited development, with a current planned buy of 552 aircraft, put the near-term future of U.S. carrier air power on a solid footing again for the first time in more than a decade. It also put the Navy on the road toward a leaner and more cost-effective air-wing structure and, by convenient happenstance, allowed for a timely test of the aircraft's combat suitability during its three-week baptism of fire in Operation Iraqi Freedom.

By 2006, all of the Navy's remaining F-14s will have been retired and replaced by a few single-seat F/A-18Es but predominantly by the two-seat F/A-18F in the fleet air defense and strike-fighter roles. Concurrently, earlier-model F/A-18As will be gradually retired from squadron service and replaced by the single-seat F/A-18E. The Navy will be able to retire the last of its Tomcats 18 months earlier than originally planned thanks to an accelerated schedule to get Super Hornets into fleet squadrons, a measure that will save the Navy about $155 million.

The existing navigation and targeting pod and laser-designator and tracker capability of the F-14's LANTIRN system and the Nitehawk pod now carried by carrier-based F/A-18s will be replaced by the ATFLIR pod that will enable the aircraft to detect, assess, classify, and track aerial and surface targets, day or night and in any weather conditions, with greater resolution and standoff range than that offered by LANTIRN. The single ATFLIR pod includes a target laser designator and laser spot tracker. It combines the navigation and targeting functions of LANTIRN, includes a visible-light camera, and frees up a weapon station. Its third-generation FLIR sensor can detect, classify, and track both aerial and surface targets at three to five times greater slant range than the Nitehawk system that it will replace.

The Super Hornet will also be equipped with the APG-79 active electronically scanned array (AESA) radar now in advanced development for installation in the Lot 26 aircraft and all that follow it. The AESA radar will be a key part of the Super Hornet's Block 2 configuration and will offer advantages not only in air-to-air scenarios but also in identifying ground targets at longer ranges and feeding that information directly to onboard weapons systems. With its automatic target acquisition and recognition feature, the new radar will have three times the detection and tracking range of the Super Hornet's current APG-73 radar and a fivefold increase in reliability, as well as the ability to jam enemy sensors while other targets are being tracked.[1] It also will allow higher-resolution ground mapping at greater ranges and will enable the Super Hornet to use synthetic aperture radar (SAR) updates to improve the accuracy of its GPS-aided bombs. The AESA radar will become operational in the Super Hornet in 2006. A concurrent addition of the helmet-mounted cueing system (HMCS) and AIM-9X infrared air-to-air missile will make the aircraft more capable in close-in engagements. (Sensors in the HMCS automatically point in whatever direction the wearer looks, with flight and target information displayed on the inside of the pilot's helmet visor.) The AIM-9X features a fifth-generation staring focal-plane array imaging seeker with near-instantaneous slew rates that will enable the missile to acquire a target at very large off-boresight angles.[2]

The two-seat F/A-18F is also slated to receive the shared reconnaissance pod (SHARP) system, which will allow the Super Hornet

---

[1] Loren B. Thompson, "The F/A-18E/F Super Hornet: Coming Soon to an Aircraft Carrier Near You," *Sea Power*, June 2000, p. 34.

[2] Richard R. Burgess, "First AIM-9X Sidewinder Delivered," *Sea Power*, June 2002, p. 45.

to collect infrared, visible, and digital SAR imagery from medium and high altitudes. By 2006, the aircraft should be able to drop 1,000 lb JDAMs. The planned buy of Super Hornets is expected to be completed by 2011, although the Navy may have to procure more should the F-35C production and deployment schedule slip (see below for more on the latter).[3]

Meeting the Needs of Electronic Warfare

After the Air Force decided to retire its 24 EF-111 Raven electronic jammers because of aging and increasingly prohibitive maintenance costs, the EA-6B Prowler became the nation's only remaining tactical electronic warfare aircraft. That made the Navy the lead service in the provision of standoff jamming, with the Air Force and Marine Corps playing a supporting role. As a result of that development, the EA-6B now fills not just a fleet commitment but a national need.

In all, Grumman Aircraft Corporation built 170 four-seat Prowlers. In 1996, 129 remained in the active inventory. The resultant shortage of jammer aircraft led to the expeditionary force of EA-6B's being classified as a low-density/high-demand (LD/HD) asset under the Department of Defense's global military force policy. Spreading EA-6Bs around in sufficient numbers to all ten of the Navy's carrier air wings has been enabled by what one expert commentator called "creative scheduling of deployment cycles, which often required the carrier-based squadrons to 'gap-fill' land-base sites during the carrier air wing turn-around cycle."[4]

The Prowler is long overdue to be replaced. It is not aerodynamically compatible with current-generation strike aircraft. Not only is it g-limited, it cannot keep up with a strike package of F/A-18s. These performance shortcomings have forced EA-6B aircrews to

---

[3] Lorenzo Cortes, "Tomcat Transition to Super Hornet Complete by Fall '06, Admiral Says," *Defense Daily*, June 16, 2004, p. 9.

[4] Commander Allan J. Assel, USN, "Airborne Electronic Attack: What's Next?" *Proceedings*, February 2001, p. 52.

devise innovative tactics, techniques, and procedures to operate effectively with strike fighters. Some mission profiles require that the Prowler get a head start on the strike package, which then complicates the problem of dealing with potential air-to-air threats, since the aircraft is vulnerable to enemy fighters and requires a fighter escort to alert and defend it.

The Improved Capability II (ICAP II) system that currently equips the bulk of the EA-6B inventory is largely limited to preemptive jamming, which demands fairly large amounts of jamming power spread over a wide frequency range and large areas of coverage. That electronic attack tactic was described by a Kosovo-experienced EA-6B crewmember as "analogous to using free-fall Mk 80-series dumb bombs in strike warfare."[5] The ICAP III system slated to be introduced in 2005 promises to provide a faster and more accurate receiver able to pinpoint enemy SAM radar frequencies and to concentrate jamming power on the desired target at the appropriate moment. As one account put it, perhaps the most important feature of ICAP III is its introduction of what has been called a "selective reactive" jamming capability. Instead of expending energy to jam the entire electromagnetic spectrum on which enemy threat radars might be operating, the new system will "automatically identify, prioritize, and jam only on those frequencies actually in use."[6] As explained in 2002 by the EA-6B program manager, ICAP III will further offer the ability to focus more concentrated jamming transmissions against specific radar frequencies, as well as to follow frequency-agile radars within their band as they switch frequencies to avoid jamming.[7] Operational assessment and carrier suitability trials of the system were completed in 2003. As for its operational promise, it has been portrayed as being to electronic attack "what precision-guided weapons are to strike warfare."[8]

---

[5] Assel, "Airborne Electronic Attack: What's Next?"

[6] Loren Thompson, "Shaping the Battlespace: The Future of Airborne Electronic Warfare," *Sea Power*, March 2000, p. 42.

[7] Richard R. Burgess, "ICAP III Prowler Begins EW Testing at Patuxent River," *Sea Power*, February 2002, p. 23.

[8] Assel, "Airborne Electronic Attack: What's Next?" p. 52.

The Navy's former Prowler program manager called the pod that currently equips the EA-6B "the preeminent SOJ [standoff jamming] system in the world today."[9]

Informal studies of EA-6B replacement options began in 1995, with the prime candidate being an evolved F/A-18F tailored to the electronic warfare mission. The current plan is for the EA-6B inventory to be replaced by 90 two-seat EA-18Gs, with the first development aircraft now slated to fly in 2006. The aircraft will have an ALQ-218 wideband receiver, ALQ-99 jamming pods, and an advanced electronic countermeasures officer's crew station in the aft cockpit. It will conduct wideband electronic support and attack, selective reactive jamming, precision threat location and targeting, and active defense suppression using the AGM-88 high-speed anti-radiation missile (HARM). It will also be able to interlink with other systems for conducting joint operations. To control program cost and risk, some anticipated capability will be deferred for later enhancement efforts, such as the integration of the joint standoff weapon (JSOW). However, the initial delivery of EA-18Gs may include the advanced anti-radiation guided missile (AARGM) with a multimode seeker that will allow the missile to continue guiding even if a targeted enemy SAM radar ceases emitting. A major advantage of the EA-18G is that it will be mission-compatible with strike fighters and will have a data-link system that will permit the transfer of digital target information between it and the strikers. The aircraft will also have a credible air-to-air capability with current missiles and the HMCS and will be less dependent on fighter escort as a result.

The EA-18G may be regarded as the bridge to the Navy's true next-generation penetrating jammer. Looking further downstream at the prospects for naval electronic warfare, integrating unmanned combat air vehicles (UCAVs) into carrier air wings as combat support jammers will not be easy, since the electronic attack mission requires no less human involvement than the strike mission. Technology will be both the driver and the limiting factor in this process. Just as strike

[9] Captain Dennis H. Krieger, USN (Ret.), "Let's Take Standoff Jamming to the Next Level," *Proceedings*, February 2001, p. 54.

UCAVs will come into their own only when technology will allow an air vehicle to conduct high-confidence bombing without a human crew on board, so will unmanned electronic attack become feasible only when technology allows it. That will be a tall order, since any such platform will need to approach well-defended enemy targets unobserved, engage any weapon system that its sensors may identify, automatically fire next-generation anti-radiation missiles at the targeted SAM radar, and still have enough remaining payload capacity to deliver precision munitions against a fixed or mobile target and enough remaining fuel to return to the carrier. Such a capability will not be at hand until many years after the EA-6B inventory is retired. Hence the interim need for the EA-18G, since even the F-35C, to say nothing of follow-on UCAV platforms, will not be mature enough to assume the electronic attack mission until 2020 at the earliest.

The Promise of JSF

After the Navy's three abortive attempts in a row during the early 1990s to join the stealth revolution with its since-cancelled A-12, AX, and A/FX programs, the service is now at the brink of acquiring its first true low-observable attack platform with the F-35C JSF that is now well into advanced flight testing and initial operational evaluation.

Ironically, the JSF started out not as a program intended to develop an aircraft for production and deployment but rather as the Department of Defense's Joint Advanced Strike Technology (JAST) demonstrator program, an effort initiated in the summer of 1993 to explore and then put on the shelf technologies that might yield affordable commonality in the next generation of Air Force, Navy, and Marine Corps strike fighters by using such common components as engines, avionics, and test and training equipment. The overriding idea for the JSF that subsequently emerged from this program was to develop three variants—one each for the Air Force, Navy, and Marine Corps—of a single-engine, single-seat aircraft to be built on a common production line.

Once the JSF program became a high-profile effort to develop a next-generation combat aircraft for the three services, the program made unprecedented use of modeling, simulation, and other analytical techniques to ensure that warfighters assessed their concepts of operations with a view toward an affordable set of requirements that would not drive costs in any one area. A core challenge here entailed aiming for the highest possible commonality, while tailoring the three variants of the aircraft to unique mission needs. One of the main initial objectives of the program was (and remains) to control costs so as to ensure a unit price sufficiently low that large numbers of the aircraft can be produced and downstream production funding cuts can be avoided. To achieve that demanding goal, the JSF program office followed a recommendation of the Packard Commission that warfighters and technologists be put together early during a combat aircraft development program to assess program cost and operational performance tradeoffs.[10] As the second JSF program director, Rear Admiral Craig Steidle, put it, "rather than define our requirements for a weapon system and then throw it to the acquisition force, we do the cost/performance tradeoffs and make the trades early so the warfighter knows what it's going to cost him.... So do you, the warfighter, know what this extra one-tenth of a g is going to cost? Do you know what this extra pound of payload is going to cost? . . . Key technologies must be adequately demonstrated before entering EMD [engineering and manufacturing development] in order to reduce development risk and control cost."[11]

In November 1996, development contracts were awarded to Boeing and Lockheed Martin, with the McDonnell Douglas submis-

---

[10] In 1985, President Ronald Reagan appointed the Blue Ribbon Commission on Defense Management, under the leadership of former Deputy Secretary of Defense David Packard, to study the U.S. defense acquisition process and to determine whether the application of private-sector methodologies and approaches might improve U.S. defense management business practices. The final report of that Commission (more commonly known as the Packard Commission) was submitted in July 1986. Its recommendations were subsequently implemented by President Reagan in National Security Decision Directive 219.

[11] "Coming Next: A Supersonic STOVL Fighter," interview with JSF Director Rear Admiral Craig E. Steidle, *Sea Power*, June 1997, pp. 9–10.

sion having failed to survive the initial cut. Boeing's X-32 and Lockheed Martin's X-35 were rolled out in the spring of 1999, and both aircraft flew for the first time in late 2000. Following a competitive demonstration-validation flyoff, the Department of Defense at the end of 2001 selected the Lockheed Martin candidate for full-scale engineering development as the tri-service F-35, with Pratt and Whitney Military Engines awarded a concurrent contract to develop the F135 engine, a production version of the JSF119 that was evolved from the engine used in the Air Force's F/A-22 Raptor and that was successfully flight-tested in the X-35.[12] (The F/A-22 has also been used by the JSF program as a technology baseline from which to develop and demonstrate common software applications, develop structural materials that will help to reduce aircraft weight and radar cross-section, and test new coatings that reduce radar reflectivity and increase the JSF's maintainability.)[13]

The diverse operational needs of the five main prospective customers (not only the Air Force, Navy, and Marine Corps but also the RAF and Royal Navy) have complicated the process of ensuring that buyers can afford the product that they seek. All prospective users agreed on the need for stealth, a precision weapons-delivery capability, the ability to share battlespace information, and minimal maintenance and logistic support needs. Beyond that, the individual customer requirements have diverged considerably. The Air Force needs a low-cost replacement for its A-10 and F-16. The Navy needs a durable, first-day-survivable stealth fighter to complement the F/A-18E/F—one that can tolerate the demanding at-sea carrier operating environment and carry a weapons load beyond the range of the F/A-18E/F. Finally, the Marine Corps and both the RAF and Royal Navy have a requirement for short takeoff and vertical landing (STOVL) performance.

---

[12] Richard R. Burgess, "Lockheed Martin Selected to Build Joint Strike Fighter," *Sea Power*, December 2001, p. 30.

[13] Brigadier General Bruce B. Byrum, USMC (Ret.), "Joint Strike Fighter: The Only Affordable Solution?" *Sea Power*, October 2000, p. 60.

In the end, the Navy's version of the JSF, the F-35C, became a modification of the Air Force's conventional takeoff and land (CTOL) configuration with a larger wing, strengthened landing gear and structure, and a lower radar cross-section that aims to offer the same first-day survivability as that now provided by the Air Force's stealthy F-117.[14] For this, the Navy will need a low-observable surface treatment that can retain its low radar cross-section properties while, at the same time, being easily maintained in the carrier operating environment, with sailors repeatedly pushing on the aircraft to move it on and off a flight-deck catapult.[15] The range/payload capability of the carrier-launched F-35C was set to be, at a minimum, compatible with that of the F/A-18E/F, which it will complement. However, the JSF will considerably overmatch that capability with a 700 nautical mile unrefueled mission radius that will restore to Navy carrier decks the deep-attack reach that was lost when the A-6E was retired. With external fuel tanks, that radius can be extended to more than 1,000 nautical miles, even when the aircraft is configured with two 2,000 lb JDAMs and a full complement of air-to-air missiles.

As for major design characteristics, the three service variants of the F-35 will have a common engine, avionics, and major subsystems. The weapons bay of the F-35C was designed to carry either two 2,000 lb JDAMs or two JSOWs, as well as two AIM-120 AMRAAM air-to-air missiles. The F-35C will offer about the same aerodynamic performance as the F/A-18E/F in most respects, such as thrust-to-weight ratio, maneuverability, top speed, range, endurance, and carrier bring-back capability. The big differences will be greater range, low observability to enemy radar, and conscious design for predictive maintenance, as well as a closer balance between capability and unit cost.

---

[14] Glenn W. Goodman, Jr., "Joint Strike Fighter: The Pentagon's JAST Program Has Progressed Rapidly and Shows Great Promise," *Armed Forces Journal International,* February 1996, p. 15.

[15] William B. Scott, "Joint Strike Fighter Balances Combat Prowess, Affordability," *Aviation Week and Space Technology*, August 3, 1998, p. 51.

Evolving Air-Wing Composition

Throughout most of the 1980s, a carrier air wing typically included as its surface-attack complement an A-6E medium-attack squadron and two A-7 light-attack squadrons. It also included two squadrons of F-14s for providing battle-group air defense and air superiority, an S-3 squadron for antisubmarine warfare, an EA-6B squadron for electronic warfare, an E-2C squadron for command and control and airborne early warning, and a helicopter squadron. The strike portion of a wing normally consisted of only 36 aircraft. By 1991, on the eve of Operation Desert Storm, the typical air wing consisted of 24 F-14s, 24 F/A-18s or A-7s, ten A-6Es (plus four KA-6D organic tankers), four EA-6Bs, ten S-3Bs, four E-2Cs, and six H-3s. That made for a total of 58 tactical aircraft, with only 34 of those strike-capable—and with only the A-6E capable of delivering LGBs without off-board target designation.

With the subsequent upgrading of the F-14 into an all-weather strike fighter, the ground-attack potential of the Navy's carrier air wings grew commensurately. At the same time, the A-6 and A-7 were gradually replaced by dual-role F/A-18s. With those two major force improvements, a typical carrier air wing by mid-1995 consisted of 20-24 F-14s, 20 24 F/A-18s, 12-14 A-6Es, four EA-6Bs, eight to ten S-3Bs, and six SH-60 helicopters. A typical strike package from such a wing would entail 30 to 40 aircraft, of which two to four divisions of four aircraft each would actually put ordnance on target in any given attack wave.

By 2003, the typical air wing consisted of ten F-14s, 36 F/A-18s, four EA-6Bs, eight S-3Bs, four E-2Cs, and seven SH-60s or HH-60s. That composition added up to only 50 tactical aircraft, yet a full 46 of which were now strike-capable—and *all* of which could deliver precision-guided weapons autonomously. The principal difference between it and earlier air wings was not so much in its platforms as in the sensors, munitions, and networks that gave the air wing its considerably enhanced force-projection capability. Instead of eight strike aircraft with unguided bombs limited to attacking only one target, for a total of merely two or three targets per attack wave, each individual strike aircraft could now routinely strike two or three DMPIs with high accuracy, for a total of 92 to 138 separate aim points for a given attack package.[16]

In recent years, the Navy has been examining ways to reduce the number of aircraft assigned to each squadron, a goal made possible by the increased reliability of today's platforms and systems. Because of the greatly increased punch of today's strike fighters armed with precision-guided munitions, carrier air wings no longer require as many aircraft as they did in years past. The ongoing neck-down strategy being pursued by the naval aviation community also aims to reduce the number of aircraft types in service and to acquire platforms that maximize commonality in mission support. The strategy's aim is to maintain overall air-wing combat capability while lessening the wing's logistics burden.[17] Tomorrow's CSGs will also typically consist of fewer surface ships and submarines, a scaled-down composition that was deemed by Admiral Michael Mullen, then Deputy CNO for Resources, Requirements and Assessments (OPNAV N8), to be "an acceptable risk when operating against transnational enemies that pose a limited at-sea threat."[18] The transfer of surface and submarine striking power from CSGs to ESGs will be offset by the dramatic increases in carrier air-wing striking power afforded by precision ordnance.[19]

By 2010, the strike component of a typical U.S. Navy carrier air wing will consist entirely of F/A-18s—20 Cs, 12 Es, and 12 Fs (constituting 44 strike fighters in all), as well as five EA-18Gs. The Super Hornet will take over the organic tanking role now fulfilled by the

---

[16] Rear Admiral Mark Fitzgerald, USN, "Operation Iraqi Freedom: The Payoff," *The Hook*, Winter 2003, p. 9.

[17] Rear Admiral Mark Fitzgerald, USN, "Order of Daedalians Brief," Riverside, Calif., June 5, 2004.

[18] Quoted in David L. Parsons, "Naval Aircraft and Weapon Developments," *Proceedings*, May 2003, p. 127. Admiral Mullen, who was the initial sponsor of this study during his tenure as N8, has since been confirmed as the next CNO.

[19] A single carrier air wing now offers the target-coverage equivalent of 4,000–5,000 TLAMs over the course of a 30-day operation.

S-3B, and the ASW role now fulfilled by the S-3 will be assumed by the SH-60 helicopter. During the same period, the Navy's helicopter inventory will be ramped down from seven types to only two—the SH-60 and MH-60. Helicopters will be provided with Hellfire missiles. They also will routinely perform vertical replenishment of carriers during ongoing air-wing operations.[20]

As the second decade of the 21st century approaches, F/A-18A+ and C units will transition to F-35Cs, EA-6Bs will be replaced by EA-18Gs, the E-2C inventory will be replaced by E-2D Advanced Hawkeyes with significantly increased airborne surveillance and battle-management capabilities, and UAVs and UCAVs will join the Navy's carrier air wings. The E-2D is now scheduled to attain initial operational capability in 2011, using a new rotating ultra-high-frequency (UHF) radar antenna and the ADS-18 electronically scanned array to detect potential threats to the carrier at greater distances than ever before, as well as in littoral areas and over land as well as over water.[21] As a result of these changes, the standard air-wing composition by 2020 will be even further simplified and streamlined. A typical mix will then be 12 F/A-18Es, 12 F/A-18Fs, and 20 F-35C Joint Strike Fighters. Air wings will include 44 strike fighters (both Super Hornets and JSFs), four UCAVs for ISR, five EA-18Gs, five E-2D Advanced Hawkeyes, and 11 MH-60 helicopters. All wings will be closely linked with a fusion of sensor images and other data that can be rapidly distributed. Such newly equipped and newly structured air wings will offer ten times the strike potential of the carrier air wings of the late 1980s, thanks to improvements in aircraft reliability, sortie-generation capability, and precision-attack capability, to say nothing of the introduction of stealth once the F-35 enters fleet service.

---

[20] Rear Admiral Thomas J. Kilcline, USN, "Navy Helicopters at the Core of Strike Group Capability," *The Hook*, Winter 2004, p. 8.

[21] Robert F. Dorr, "Change in the Air: What's Waiting in the Navy's Wings?" *Armed Forces Journal*, May 2005, p. 36. For more on the various needs that led to this impending initiative, see Obaid Younossi and others, *The Eyes of the Fleet: An Analysis of the E-2C Aircraft Acquisition Options*, Santa Monica, Calif.: RAND Corporation, MR-1517-NAVY, 2002.

Besides increased aircraft reliability and greatly improved munitions accuracy, another contributor to the strengthened capability of the Navy's carrier air wings has been the steady integration of Marine Corps F/A-18C strike-fighter squadrons into the wings' organizational makeup over the past decade. Although it has since proven to be highly productive on operational and tactical grounds, this integration was at first driven primarily by cost considerations, in view of the approaching post-cold-war bow wave for tactical aircraft. Budget constraints and the post-cold-war force drawdown dictated a partial integration in 1997, when four Marine F/A-18C squadrons began making regular deployments with four of the Navy's ten carrier air wings.

Since then, this integration has been further institutionalized, to a point where ten Marine Corps strike-fighter squadrons are now slated to join the Navy's air wings, with one squadron formally assigned to each wing. In accordance with that plan, four additional Marine Corps F/A-18C squadrons will assume Navy air-wing commitments by 2007. In a reciprocal arrangement, the Navy will put three of its Hornet squadrons into Marine air wings to support the Marine wings' unit deployment program. Sea-based Marine Corps squadrons will be reduced from 12 to ten aircraft, as will all Navy F/A-18A+ and C squadrons. With two squadrons per air wing, that will mean a decline of four strike fighters in each wing. (Shore-based Marine Corps strike-fighter squadrons will remain at 12 aircraft per squadron.) The first Marine carrier air wing commander has already been selected, as has a Navy captain to command a Marine Corps expeditionary air group.

In addition to the manned aircraft air-wing complements noted above, the Navy also is looking ahead toward the eventual introduction of UAVs into its air wings, initially to provide surveillance and reconnaissance for CSGs and, later as the requisite technology further matures, to conduct actual suppression of enemy air defenses (SEAD) and strike operations in the most heavily defended target areas. For the near term, the SHARP pod on the two-seat F/A-18F should offer an adequate solution to the need for real-time ISR. The longer-range solution, however, is seen as lying in UAVs that will offer longer dwell time over potential targets and other areas of interest.

The Navy's plan for acquiring UAVs anticipates three levels of operational use—strategic, operational, and tactical. The uppermost of these echelons will be the domain of a yet-to-be-chosen broad-area maritime surveillance (BAMS) aircraft that will survey a broad expanse of ocean and land space from very high altitude. These land-based aircraft will be large, on the order of the Air Force's RQ-4 Global Hawk that fulfilled such vital ISR functions in Operations Enduring Freedom and Iraqi Freedom. Such UAVs will conduct standoff surveillance and collect electronic and signals intelligence that can be relayed in real time to the area commander and other interested fusion centers. They also may serve, in effect, as what one account characterized as "low-hanging satellites," equipped with broad-band transponders so that they could be used as communications links to relay signals between aircraft and ships that are over the horizon from one another and accordingly could not communicate otherwise. In so doing, such UAVs will handle part of the job currently performed by the Navy's P-3 Orion, with the remainder being shouldered by a follow-on multimission maritime aircraft (MMA).[22]

Later, stealthy unmanned *combat* aerial vehicles designed expressly to perform carrier-based strike operations will also be introduced into fleet service. The stimulus for this second area of UAV inquiry by the Navy has stemmed in part from a recent House Armed Services Committee insistence that UCAVs constitute a third of all U.S. deep-strike assets by 2010 as a precondition for further program funding. The Navy plans to have a small squadron of UCAVs operate off carrier decks by 2015. However, as Admiral Fitzgerald has cautioned, the UCAV will have to "earn its way onto the ship."[23] Unmanned aircraft will first have to prove that they are sufficiently reliable to be safely launched from and recovered aboard carriers. A related issue, not just for the Navy but for all of the services, concerns whether UCAVs can perform their expected missions better than

[22] David Brown, "BAMS, Eagle Eyes, and Dragon Eyes," *Sea Power*, April 2003, p. 66.

[23] Lorenzo Cortes, "Tomcat Transition to Super Hornet Complete by Fall '06, Admiral Says," *Defense Daily*, June 16, 2004, p. 9.

manned aircraft in high-threat and high-risk environments. Under the aegis of the Joint Unmanned Combat Aerial System (J-UCAS) program, the Defense Advanced Research Projects Agency is now funding the construction of UCAVs and plans to fly them from carrier-like decks in tests. Ultimately, however, UAVs and UCAVs will surely constitute a vital organic part of the Navy's carrier air-wing structure. Currently, the Navy plans to have the first UCAVs enter fleet service in 2015 and to eventually incorporate six to nine UCAVs into each carrier air wing.[24]

Toward a More Effectively Linked Force

The Navy's first employment of concepts and capabilities for network-centric operations in preparation for actual combat occurred on the eve of the terrorist attacks of September 11, 2001. As USS *Carl Vinson* was preparing to relieve *Enterprise* in CENTCOM's AOR the day before, senior leaders in both ships had already come to realize that traditional message traffic was too slow, layered, and rigid. Each of the two battle groups had come from a separate U.S. coast, and each had its own unique procedures that it had developed during its respective predeployment workup. Fortunately, however, both groups had access to the same link picture, thanks to a grassroots innovation called "knowledge web" (Kweb) that used SIPRNET as a medium for creating a battle-group Web site.

This "knowledge web" was tested in *Carl Vinson* and Carrier Group Three under the command of Rear Admiral Thomas Zelibor. All ships in the battle group that were tied to SIPRNET were able to access the Web, although some lacked sufficient bandwidth to surf the Web extensively. Kweb included an extensive series of pages that

---

[24] Ronald O'Rourke, "Unmanned Vehicles for U.S. Naval Forces: Issues for Congress," *CRS Report for Congress,* Order Code RS21294, Washington, D.C.: Congressional Research Service, the Library of Congress, May 30, 2003, p. 3. For more on likely initial concepts of operations for the Navy's first generation of carrier-based UCAVs, see James Perry and Adam Siegel, "J-UCAS: Enhancing Naval Aviation's Contribution to National Security," *Wings of Gold,* Spring 2005, pp. 12–14.

showed real-time air, naval, and ground activity during Operation Enduring Freedom. It also featured a current situation-report page that provided the battle group with a common operating picture. That eliminated any need for Powerpoint intelligence briefings and a daily commander's meeting with his staff. For example, the air wing commander aboard *Carl Vinson* did not have to tell Admiral Zelibor that three of his aircraft were down for maintenance, since the admiral had already noted those facts online several hours earlier because the air wing's maintenance officer had already updated that section of his page.[25] As Admiral Fitzgerald said later of such early-generation network-centric applications, the carrier air contribution to Enduring Freedom revealed "the tip of the revolution that's continuing" in naval aviation. The first revolution was in the realm of precision weapons and precision targeting. The second, the admiral said, will entail "going from the analog to the digital age in communications architecture."[26]

The multifunction information distribution system (MIDS), a nodeless and secure Link 16-based jam-resistant tactical data link, also made a major difference in enabling enhanced interoperability with other joint and multinational platforms equipped with that capability. Now in the fleet and with more than a thousand Link 16 terminals in the four services, it was a major contributor to the continuing transition from analog to digital warfighting and paved the way for the next step in network-centric operations.[27] Admiral Fitzgerald cited this experience as further proof that the Navy must strive to become a "fully linked force," with continued investment in MIDS/Link 16 and the Joint Fires Network (JFN) to move target

---

[25] For further details, see Captain Eileen F. MacKrell, USN, "Network-Centric Intelligence Works," *Proceedings*, July 2003, pp. 44–48.

[26] "Fitzgerald: Recapitalization Poses Challenge for Naval Air," *Sea Power*, March 2004, p. 28.

[27] Captain David C. Hardesty, USN, "Fix Net Centric for the Operators," *Proceedings*, September 2003, p. 69.

information more quickly from sensors into the cockpits of strike fighters.[28]

Looking to the future, the Navy is clearly evolving from being a platform-centric to a network-centric force. Future CSGs and their embarked air wings, furthermore, will increasingly aim for effects-based operations rather than classic attrition operations. Toward both ends, the Navy now uses the overarching term FORCENet to characterize its ongoing process for applying network-centric theory. Because it is not a discrete system or array of connected hardware so much as an organizing concept for getting the greatest combat leverage out of the digital data stream, FORCENet is hard to describe in a nutshell. At its core, however, it is a systematic approach aimed at improving combat decisionmaking at all levels by creating a seamless grid of interconnected sensors, weapons, individuals, and command and control mechanisms that are accessible to all elements of the fleet and are studiously designed to enhance their ability to sense, locate, communicate, attack, and assess. As described in a recent report to Congress by the Secretary of the Navy, "FORCENet is not an acquisition program; rather, it is an enterprise alignment and integration initiative to serve as a change agent and an engine for innovation, potentially touching every naval program"[29]

One of the principal payoff areas of network-centric operations, time-sensitive targeting, seeks to reduce the time between the moment a target is detected and the moment it is attacked. Its goal is to increase the number of DMPIs a CSG can engage daily from a few hundred to more than five times that many as naval strike aircraft increase their payload of miniaturized weapons. A kill chain measured in single-digit minutes (and ultimately as little as seconds) can be achieved by improvements in the sensor-to-shooter interface. The time required to elicit a target approval and subsequent decision to attack, however, will continue to be the key rate-limiter. Better intel-

---

[28] Sandra I. Erwin, "Iraqi Freedom Tests Naval Aviation's Flexibility," *The Hook*, Summer 2003, p. 65.

[29] Rear Admiral Steven J. Tomaszeski, USN, "Heart of FORCENet: Sensor Grid, Advanced Command and Control," *Sea Power*, March 2004, p. 16.

ligence preparation of the battlefield and what the Air Force calls predictive battlespace awareness will concentrate on most likely enemy operating areas and will dynamically cross-cue sensors to fuse as rapidly as possible all pertinent target data collected by networked sensors.

The watchword in all of this will be persistent, long-dwell ISR to deny the enemy a day, night, or weather sanctuary. Its goal will be to operate inside any enemy's decision timeline and thereby foreclose his tactical options. This will come both from enhancing legacy networks and acquiring and fielding new ones. As Admiral Fitzgerald has suggested, "where we are trying to get to ultimately is . . . a fully linked force. We want to have 'Internet-in-the-cockpit' capability. We want to have the ability for a pilot flying in harm's way to call up and say, 'OK, what is my target looking like right now? What is the latest that I am getting out of that Predator or that Global Hawk right now? What is the best information that I have got [regarding] friendly forces on the ground?'"[30] The eventual goal of FORCENet is an ever-tighter fusion of data networks, sensors, platforms, and weapons aimed at producing what the naval aviation vision statement has called "a single, comprehensive maritime network for the battlespace" made possible by a seamless connection of all naval and joint assets in a theater of operations through their sensors, networks, decision aids, weapons, and supporting systems.[31]

---

[30] Hunter Keeter, "Navy's Lessons from Afghanistan, Iraq Include Networked Tactical Aircraft," *Defense Daily*, May 6, 2003, p. 3.

[31] Keeter, "Navy's Lessons from Afghanistan, Iraq Include Networked Tactical Aircraft," p. 6.

CHAPTER SEVEN

Conclusions

Operations Enduring Freedom and Iraqi Freedom showed that the Navy's aircraft carriers no longer operate as individual and autonomous air-wing platforms but rather as a surged and massed force capable of generating and sustaining however many consistently effective sorties over time that a CFACC might need to meet his assigned campaign goals. This performance was a direct outgrowth of the Navy's having corrected many of its operational, doctrinal, and force capability deficiencies that were first spotlighted during Operation Desert Storm in 1991. Unlike the relatively short sorties flown during the punitive strikes that had been launched earlier against Lebanon in 1983 and Libya in 1986 and in such subsequent contingency responses during the 1990s as Operations Deliberate Force, Desert Fox, and Allied Force, these were multicycle missions that lasted for as long as ten hours and that ranged deep beyond littoral reaches into the heart of Afghanistan and Iraq, the first of which was landlocked in the remotest part of Southwest Asia. In each case, they showed that U.S. carrier air power can carry out sustained strikes against targets several hundred miles inland and, with the support of nonorganic tanking and needed underway replenishment at regular intervals, can do so for weeks or even months should operational needs require.

To be sure, each of the U.S. armed services in those two wars brought a needed comparative advantage to the fight. For example, in the case of Enduring Freedom, although Air Force bombers flew only around 10 percent of the total number of combat sorties, they dropped roughly 80 percent of the ordnance, including the preponderant number of satellite-aided JDAMs. Nevertheless, what matters most here is that although the Navy indeed needed the support of Air Force and allied long-range tankers to be mission-effective, it also turned in a professional performance in a part of the world where the Air Force both lacked the needed access to operate its fighters most efficiently and remained limited in the number of combat sorties it could generate even after it finally achieved its needed access. The reason for that was distances from forward land bases to target that demanded fighter missions lasting as long as 15 hours, which were unsustainable by the Air Force over the long haul.

By every measure that matters, U.S. naval aviation acquitted itself well during the first two wars of the 21st century. The previous inability of the carrier force to project credible and sustained combat power at great distances was a persistent and common theme of critiques of sea-based strike aviation in the roles and missions debates throughout the 1990s. The substantial contribution of naval strike and combat-support assets to those two subsequent wars in close succession did much to dispel that widespread and deeply held conviction.

That said, naval aviation by no means faces smooth sailing when it comes to the looming budget battles on the domestic front. To note but one example, between 1995 and 2000, its aircraft force-structure numbers finally stabilized at 46–47 strike aircraft per air wing. Yet at the current and projected funding rate, the Navy may not be able to attain the desired number of 44 strike fighters for each carrier plus the required number of additional strike fighters for the Marine Corps. At the start of 2003, the Navy and Marine Corps possessed nearly 3,800 aircraft out of a total stated requirement for 4,260. That was close to 90 percent of the mandated total.[1] However, the average age of those aircraft makes that seemingly close percentage figure problematic because of their ever-decreasing maintainability and reliability. Thus, although the number of authorized carriers re-

---

[1] Captain Mike Spence, USN, "Naval Aviation Is Behind the Power Curve," *Proceedings,* February 2003, p. 48.

mains a critically important concern, it is not the most burning issue facing naval aviation today. A more overarching concern is whether the Navy will be able to acquire and retain enough modern and capable aircraft to keep its authorized number of carrier decks filled.

All of that notwithstanding, the United States remains the only country in the world capable of deploying a carrier-based strike force of any significant size. Today, deep-strike carrier aviation is not only a natural concomitant of the nation's status as the world's sole surviving superpower, it also is the one outstanding feature that distinguishes the U.S. Navy unequivocally from all other naval forces around the world. The *Nimitz*-class carrier has often been described as four and a half acres of sovereign U.S. territory that can go anywhere the nation's leaders may desire to send it without needing a permission slip. For years, that characterization was routinely dismissed by critics of carrier air power as a mere slogan that overlooked the fact that a carrier can be in only one place at a time, irrespective of where a need for it might suddenly arise. Such criticism may have had merit throughout most of the cold war years, when the Navy typically kept only two or three carrier battle groups deployed at sea at any given time, while the others and their associated air wings remained at home in various states of maintenance and requalification training that rendered them undeployable on short notice. That is no longer true, however, in today's world of constant carrier surge capability under the FRP.

As for the value offered by carrier air power to the nation's combatant commanders around the world, many observers maintained for years that the continued absence of any plan for a new medium-attack aircraft to replace the A-12 was a tacit indicator that the Navy had, in effect, opted out of the business of deep-strike operations. Yet the sustained daily generation of mission-effective sorties over more than a three-month period of time that was demonstrated by the four air wings that took part in Enduring Freedom, as well as the extended distances to target that uniformly characterized their missions, would suggest that the exact opposite of that intimation is now the case. If anything, the very *essence* of U.S. carrier aviation today, with the help of nonorganic Air Force and allied tanker support, has become deep and sustained attack from the sea—as the principal player when nearby land bases are not available, as was the case in Enduring Freedom, and as a still-needed and much-welcomed equal contributor to joint and combined operations when adjacent shore bases *are* accessible, as was the case in Iraqi Freedom. Not only that, once the F-35C achieves initial operational capability toward the end of this decade, the Navy will finally have acquired a dedicated stealth attack platform with a degree of reach comparable to that of the venerable A-6 medium-attack bomber that last flew in December 1996 after more than three decades of storied service with the fleet.

Admittedly, the well-worn criticism of the aircraft carrier by some as a "self-licking ice cream cone" with respect to its continuing need to protect itself as a first priority remains valid enough as far as it goes. Yet quite apart from the fact that the imperatives of base defense and force protection are by no means unique to aircraft carriers, the Navy's performance in Operations Enduring Freedom and Iraqi Freedom showed clearly how far the nation's carrier force has advanced since Desert Storm in terms of providing around-the-clock target coverage, consistently accurate target attack, and—for the first time in its history—multiple successful target attacks per sortie. In 2011, U.S. naval aviation will celebrate its 100th anniversary. When that milestone is reached, American carrier air power's classic roles and missions will not have changed greatly from those of the 20th century. Yet the nation's carrier strike groups will have taken on substantially increased combat leverage with the completion of the Super Hornet acquisition, the advent of the F-35C and EA-18G, and the prospective introduction of UAVs and UCAVs into the Navy's carrier air-wing complements.

Now that the Soviet air and naval threat, against which the American carrier inventory was configured throughout the many years of the cold war, has finally disappeared, there can be no doubt that that inventory will come under ever-increasing pressure to show its relevance and utility in nontraditional applications, as USS *Kitty Hawk* demonstrated during Enduring Freedom when she performed as a dedicated staging base for SOF helicopters and as USS *Abraham Lincoln* subsequently performed, in the most extensive U.S. military operation in Southeast Asia since the Vietnam War, in providing tsunami relief in early 2005 to the hardest-hit areas of Indonesia. Yet as the American defense establishment sets about reconfiguring itself for the most likely wars of the near-term future, its leaders would do well to remember the injunction of the late Ambassador Robert Komer that in preparing for such wars, they should not forget to hedge also against the one they could lose. As they anticipate and plan against such worst-case contingencies and the lesser included cases that will inevitably occur along the way, the nation's leaders can draw substantial insurance and actual combat power, in times of need, from the proven reach, mobility, access, precision-strike capability, offensive persistence, and related advantages that are quintessentially offered by the Navy's carrier force.

Bibliography

Books

Franks, General Tommy, with Malcolm McConnell, *American Soldier,* New York: Regan Books, 2004.

Keegan, John, *The Iraq War,* New York: Alfred A. Knopf, 2004.

Marolda, Edward J., and Robert J. Schneider, Jr., *Sword and Shield: The United States Navy and the Persian Gulf War,* Annapolis, Md.: Naval Institute Press, 1998.

Owens, Admiral William A. (Ret.), *High Seas: The Naval Passage to an Uncharted World,* Annapolis, Md: Naval Institute Press, 1995.

Purdum, Todd S., *A Time of Our Choosing: America's War in Iraq,* New York: Times Books, 2003.

Winnefeld, James A., and Dana J. Johnson, *Joint Air Operations: Pursuit of Unity in Command and Control, 1942–1991,* Annapolis, Md.: Naval Institute Press, 1993.

Woodward, Bob, *Plan of Attack,* New York: Simon and Schuster, 2004.

Monographs and Reports

Defense Science Board Task Force on Future of the Aircraft Carrier, Washington, D.C.: Office of the Under Secretary of Defense for Acquisition, Technology, and Logistics, October 2002.

Jewell, Angelyn, Maureen A. Wigge, and others, *USS Nimitz and Carrier Air Wing Nine Surge Demonstration*, Alexandria, Va.: Center for Naval Analyses, CRM 97-111.10, April 1998.

Weldon, William F., and others, *CVX Flexibility*, Washington, D.C.: Office of the Assistant Secretary of the Navy for Research, Development and Acquisition, Naval Research Advisory Committee Report NRAC 97-1, October 1997.

Whiteway, Roger N., and Thomas Vance, *The Military Effectiveness of CVX Propulsion Alternatives*, McLean, Va.: Science Applications International Corporation, September 1998.

Younossi, Obaid, and others, *The Eyes of the Fleet: An Analysis of the E-2C Aircraft Acquisition Options*, Santa Monica, Calif.: RAND Corporation, MR-1517-NAVY, 2002.

Journal and Periodical Articles

Assel, Commander Allan J., USN, "Airborne Electronic Attack: What's Next?" *Proceedings*, February 2001.

Barry, Major General John L., USAF, and James Blaker, "After the Storm: The Growing Convergence of the Air Force and Navy," *Naval War College Review*, Autumn 2001.

Bohmfalk, Christian, and Jonathan Block, "Roosevelt Carrier Battle Group Scheduled to Deploy Wednesday," *Inside the Navy*, September 17, 2001.

Braswell, Commander Donald H., "Carrier Strike Groups Need a Combat Surge Capability," *Proceedings*, September 2004.

Brown, David, "BAMS, Eagle Eyes, and Dragon Eyes," *Sea Power*, April 2003.

Burgess, Richard R., "Super Hornet Tallies 1000 Hours; AMRAAM Shot Successful," *Sea Power*, July 1997.

__________, "Newport News Gears Up for CVN Refueling," *Sea Power*, September 1997.

__________, "Lockheed Martin Selected to Build Joint Strike Fighter," *Sea Power*, December 2001.

__________, "ICAP III Prowler Begins EW Testing at Patuxent River," *Sea Power*, February 2002.

__________, "First AIM-9X Sidewinder Delivered," *Sea Power*, June 2002.

__________, "Sea Services Strike for Iraqi Freedom," *Sea Power*, April 2003.

Byrum, Brigadier General Bruce B., USMC (Ret.), "Joint Strike Fighter: The Only Affordable Solution?" *Sea Power*, October 2000.

Castelli, Christopher J., "Enduring Freedom's Not-So-Secret Weapon: Navy F-14 LANTIRN Pods," *Inside the Navy*, April 29, 2002.

Clark, Admiral Vern, USN, "Persistent Combat Power," *Proceedings*, May 2003.

"Coming Next: A Supersonic STOVL Fighter," Interview with JSF Director Rear Admiral Craig E. Steidle, *Sea Power*, June 1997.

Cortes, Lorenzo, "Tomcat Transition to Super Hornet Complete by Fall '06, Admiral Says," *Defense Daily*, June 16, 2004.

Dorr, Robert F., "Change in the Air: What's Waiting in the Navy's Wings?" *Armed Forces Journal*, May 2005.

Erwin, Sandra I., "Iraqi Freedom Tests Naval Aviation's Flexibility," *The Hook*, Summer 2003.

__________, "Naval Aviators Experience Success in Iraq, But Worry About the Future," *The Hook*, Fall 2003.

Fitzgerald, Rear Admiral Mark, USN, "Operation Iraqi Freedom: The Payoff," *The Hook*, Winter 2003.

"Fitzgerald: Recapitalization Poses Challenge for Naval Air," *Sea Power*, March 2004.

Friedman, Norman, "World Naval Developments," *Proceedings*, July 2003.

Fulghum, David A., "Tanker Puzzle: Aggressive Tactics, Shrinking Tanker Force Challenge Both Planners and Aircrews," *Aviation Week and Space Technology*, April 14, 2003.

__________, "New Bag of Tricks: As Stealth Aircraft and Northern Watch Units Head Home, Details of the Coalition's Use of Air Power Are Revealed," *Aviation Week and Space Technology*, April 21, 2003.

Goodman, Glenn W., Jr., "Joint Strike Fighter: The Pentagon's JAST Program Has Progressed Rapidly and Shows Great Promise," *Armed Forces Journal International,* February 1996.

Grant, Rebecca, "The Carrier Myth," *Air Force Magazine*, March 1999.

Hardesty, Captain David C., USN, "Fix Net Centric for the Operators," *Proceedings*, September 2003.

Harrison, Lieutenant Commander Richard K., USN, "TacAir Trumps UAVs in Iraq," *Proceedings*, November 2003.

Keating, Vice Admiral Timothy J., "This Was a Different War," *Proceedings*, June 2003.

Keeter, Hunter, "Navy's Lessons from Afghanistan, Iraq Include Networked Tactical Aircraft," *Defense Daily*, May 6, 2003.

__________, "Tanking, 'Bring-Back' Highlights of Super Hornet's Performance," *Defense Daily*, June 2, 2003.

Kilcline, Rear Admiral Thomas J., USN, "Navy Helicopters at the Core of Strike Group Capability," *The Hook*, Winter 2004.

Krieger, Captain Dennis H., USN (Ret.), "Let's Take Standoff Jamming to the Next Level," *Proceedings*, February 2001.

Lazisky, Richard, "CVX: Evolution of a Carrier," *Sea Power*, October 1998.

MacKrell, Captain Eileen F., USN, "Network-Centric Intelligence Works," *Proceedings*, July 2003.

Malone, Vice Admiral Michael, USN, "From Readiness at Any Cost to Cost-Wise Readiness," *The Hook*, Summer 2004.

Mazach, Vice Admiral John, USN (Ret.), and Adam Siegel, "CVN-21 and Sea Power 21: A Program Aligned with the Navy's Transformational Vision," *Wings of Gold*, December 2003.

Nathman, Rear Admiral John B., "A Revolution in Strike Warfare," *Sea Power*, October 1999.

Nathman, Vice Admiral John B., USN, "We Were Great: Navy Air in Afghanistan," *Proceedings*, March 2002.

Natter, Admiral Robert J., "Creating a Surge Ready Force," *Proceedings*, September 2003.

Nichols, Rear Admiral David C., Jr., USN, "Reflections on Iraqi Freedom," *The Hook,* Fall 2003.

"Notes from the Precision Strike Conference, Fort Belvoir, Virginia, April 16," *Inside the Navy,* April 29, 2002.

Nutwell, Captain Robert M., USN, "Leaner with Marines," *Proceedings,* January 1991.

O'Connor, Commander Cathal, USN, "Surge-Ready Strike Group Offers Lessons," *Proceedings,* December 2003.

Oliver, Commander J. D., "Use the Carriers or Lose Them," *Proceedings,* September 1993.

Owens, Vice Admiral William, USN, "The Quest for Consensus," *Proceedings,* May 1994.

Parsons, David L., "Naval Aircraft and Weapon Developments," *Proceedings,* May 2003.

Perin, David A., "Are Big Decks Still the Answer?" *Proceedings*, June 2001.

Perry, James, and Adam Siegel, "J-UCAS: Enhancing Naval Aviation's Contribution to National Security," *Wings of Gold,* Spring 2005.

Scott, William B., "Joint Strike Fighter Balances Combat Prowess, Affordability," *Aviation Week and Space Technology,* August 3, 1998.

Sherman, Jason, "Defense Department Eyes Further Cuts to Aircraft Carrier Fleet," *Inside the Pentagon,* April 21, 2005.

Siegel, Adam, and Robert M. Schatzel, "CVN-21 and HSI [Human Systems Integration]," *Wings of Gold,* April 2004.

Spence, Captain Mike, USN, "Naval Aviation Is Behind the Power Curve," *Proceedings,* February 2003.

Thompson, Loren B., "Shaping the Battlespace: The Future of Airborne Electronic Warfare," *Sea Power,* March 2000.

__________, "The F/A-18E/F Super Hornet: Coming Soon to an Aircraft Carrier Near You," *Sea Power,* June 2000.

__________, "Transformation vs. Tradition," *Sea Power,* June 2002.

Tomaszeski, Rear Admiral Steven J., USN, "Heart of FORCENet: Sensor Grid, Advanced Command and Control," *Sea Power,* March 2004.

Towell, Pat, "Crusader May Be Precursor to More Defense Cuts," *Congressional Quarterly Weekly*, July 20, 2002.

Truver, Scott C., "Today's Carrier Force—and Tomorrow's," *Sea Power*, June 2001.

_________, "The U.S. Navy in Review," *Proceedings*, May 2003.

Vego, Milan, "What Can We Learn from Enduring Freedom?" *Proceedings*, July 2002.

Wages, Colonel Brian E., USAF (Ret.), "Circle the Carriers: Why Does 'Virtual Presence' Scare the Navy," *Armed Forces Journal International*, July 1995.

Wall, Robert, "Targeting, Weapon Supply Encumber Air Campaign," *Aviation Week and Space Technology*, October 22, 2001.

_________, "EA-6B Crews Recast Their Infowar Role," *Aviation Week and Space Technology*, November 19, 2001.

_________, "F-14s Add Missions in Anti-Taliban Effort," *Aviation Week and Space Technology*, November 19, 2001.

_________, "E-War Ramps Up: EA-6B Prowler to Resume Traditional Radar-Jamming Role if Iraqi Conflict Escalates," *Aviation Week and Space Technology*, March 17, 2003.

Wolfe, Frank, "Navy F-14s Able to Transmit, Receive Imagery from Green Berets in Afghanistan," *Defense Daily*, August 1, 2002.

Woods, Randy, "Prowler, Hawkeye Pilots See Roles Expanding in Enduring Freedom," *Inside the Navy*, May 6, 2002.

Newspaper Articles

Arkin, William M., "A Week of Air War," *Washington Post*, October 14, 2001.

Baker, Peter, "Marine Predicts Brief Bombing, Then Land Assault," *Washington Post*, March 17, 2003.

Baker, Peter, and Rajiv Chandrasekaran, "Republican Guard Units Move South from Baghdad, Hit by U.S. Forces," *Washington Post*, March 27, 2003.

Boudreaux, Richard, "Two Errant Missiles Fall in Turkey," *Los Angeles Times*, March 24, 2003.

Burns, Robert, "U.S. Gulf Force Nears 300,000 as Commander, Bush Consult," *Philadelphia Inquirer*, March 5, 2003.

Dao, James, "More U.S. Troops in bin Laden Hunt; Hideouts Bombed," *New York Times,* November 19, 2001.

Eisman, Dale, "Senator Warner Opens Battle to Maintain Norfolk Fleet," *Norfolk Virginian-Pilot*, February 11, 2005.

Frantz, Douglas, "Hundreds of al Qaeda Fighters Slip into Pakistan," *New York Times,* December 19, 2001.

Gellman, Barton, and Dana Priest, "CIA Had Fix on Hussein," *Washington Post,* March 20, 2003.

Gertz, Bill, and Rowan Scarborough, "Pentagon to Send Fourth Carrier to Afghanistan," *Washington Times*, November 8, 2001.

Gordon, Michael R., "The Goal Is Baghdad, But at What Cost?" *New York Times*, March 25, 2003.

Graham, Bradley, "U.S. Military Feels Strain of Buildup," *Washington Post,* February 5, 1998.

Jaffe, Greg, "U.S. Armed Forces Are Put on the Highest State of Alert," *Wall Street Journal,* September 12, 2001.

_________, "Plan Is to Cut Off Top Officers While Allies Strike Air Defenses," *Wall Street Journal*, March 20, 2003.

Layton, Lyndsey, "Building Bombs Aboard the Abraham Lincoln," *Washington Post*, March 14, 2003.

Loeb, Vernon, "Afghan War Is a Lab for U.S. Innovation," *Washington Post*, March 26, 2002.

Lynch, David J., and John Diamond, "U.S., British Forces Are 'Ready Today' for Invasion," *USA Today*, March 17, 2003.

Milbank, Dana, and Mike Allen, "President Tells Hussein to Leave Iraq Within 48 Hours or Face Invasion," *Washington Post*, March 18, 2003.

Moniz, Dave, and John Diamond, "Attack on Guard May Be Days Away," *USA Today*, March 31, 2003.

Pan, Philip P., "Turkish Leader Makes Request on Airspace," *Washington Post*, March 20, 2003.

Ricks, Thomas E., "Unfolding Battle Will Determine Length of War," *Washington Post*, March 25, 2003.

Ricks, Thomas E., and Vernon Loeb, "Initial Aim Is Hitting Taliban Defenses," *Washington Post*, October 8, 2001.

Robbins, Carla Anne, Greg Jaffe, and Dan Morse, "U.S. Aims at Psychological Front, Hoping Show of Force Ends War," *Wall Street Journal*, April 7, 2003.

Sanger, David E., with John F. Burns, "Bush Orders an Assault and Says America Will Disarm Foe," *New York Times*, March 20, 2003.

Scarborough, Rowan, "U.S. Splits Afghanistan into 'Engagement Zones,'" *Washington Times*, October 18, 2001.

__________, "Air Force Slow to Transfer Special Bomb Kits to Navy," *Washington Times*, November 7, 2001.

Schrader, Esther, "U.S. Keeps Pressure on al Qaeda," *Los Angeles Times*, January 8, 2002.

Shadid, Anthony, "In Shift, War Targets Communications Facilities," *Washington Post*, April 1, 2003.

Vogel, Steve "Gas Stations in the Sky Extend Fighters' Reach," *Washington Post*, November 1, 2001.

Williams, Carol J., "Navy Does Battle with Sandstorms on the Sea," *Los Angeles Times*, March 27, 2003.

__________, "Super Hornet Creates a Buzz in the Gulf," *Los Angeles Times*, April 1, 2003.

Briefings

Fitzgerald, Rear Admiral Mark, USN, "Order of Daedalians Brief," Riverside, Calif., June 5, 2004.

"Fleet Response Plan: Postured to Deter, Poised to Strike," briefing prepared by OPNAV N3/5 (Deep Blue), Washington, D.C.: Office of the Chief of Naval Operations, March 2004.

Green, Vice Admiral Kevin, USN, Deputy CNO for Plans, Policy, and Operations (N3/N5), "Operation Iraqi Freedom," briefing to Representative Roscoe Bartlett, Washington, D.C., n.d.

Moffit, Rear Admiral Matthew, USN, Commander, Naval Strike and Air Warfare Center, "Naval Aviation 2010–2020: A Decade of Transition," briefing to the Navy TACAIR Symposium 2004, n.d.

Schatzel, Robert M., *CVN-21 Program Overview*, Newport News, Va.: Northrop Grumman Corporation, May 18, 2005.

Ward, Robert W., Allen Hjelmfelt, Carter Malkasian, Daniel Roek, John Tand, and Daniel Whiteneck, "Operation Iraqi Freedom: CVW Fire Support to Ground Forces," undated briefing charts, Fairfax, Va.: Center for Naval Analyses.

Miscellaneous

Capaccio, Tony, "Sixty Percent of Bombs Dropped on Afghanistan Precision-Guided," Bloomberg.com, November 20, 2001.

Coordinated Fleet-Wide Message on FRP from the Commander, U.S. Fleet Forces Command and Commander, U.S. Pacific Fleet, April 24, 2004.

Moseley, Lieutenant General T. Michael, USAF, *Operation Iraqi Freedom—By the Numbers*, Shaw AFB, S.C.: Assessment and Analysis Division, Headquarters U.S. Central Command Air Forces, April 30, 2003.

O'Rourke, Ronald, "Navy CVNX Aircraft Carrier Program: Background and Issues for Congress," *CRS Report for Congress*, Order Code RS20643, Washington, D.C.: Congressional Research Service, the Library of Congress, November 14, 2002.

__________, "Unmanned Vehicles for U.S. Naval Forces: Issues for Congress," *CRS Report for Congress*, Order Code RS21294, Washington, D.C.: Congressional Research Service, the Library of Congress, May 30, 2003.

Wood, David, "New Workhorse of U.S. Military: A Bomb with Devastating Effects," Newhouse.com, March 13, 2003.